Nosey Beast

Natural History of the Coatis

Nosey Beast

Natural History of the Coatis

Christine C. Hass

Wild Mountain Echoes

Front cover photo by Christine Hass; back cover photo by Ben Hirsch. Other photos and illustrations by the author except where noted.

Published by Wild Mountain Echoes
Carson City, Nevada
https://www.wildmountainechoes.com/books

First Edition

Hass, Christine C., 1959-
      Nosey Beast. Natural History of the Coatis/Christine C. Hass
      Includes bibliographic references and index.
      ISBN 978-1-7366063-0-8 (paper)
      ISBN 978-1-7366063-1-5 (cloth)
      ISBN 978-1-7366063-2-2 (ebook)
      Library of Congress Control Number: 2021903544
      1. Coati - Behavior. 2. Coati - Ecology. 3. Mammals - Ecology. 4. Social behavior in animals. 5. Wildlife Conservation. I. Title.

Dedicated to the memories of Mike Seidman

and Sheridan Stone

Contents

Preface   ix

Part One Introduction

  1 What is a Coati?   1
  2 Why Coatis? Why Arizona?   19
  3 How to Catch a Coati   31

Part Two Natural History

  4 Birth   43
  5 Timing of Reproduction   55
  6 The Social Carnivoran   67
  7 Benefits of Group Living   83
  8 Costs of Sociality   99
  9 Perception and Communication   115
  10 Keeping the Group Together   131
  11 Settling Disputes   145
  12 Sending the Alarm   153
  13 What to Eat   165
  14 Where to Live   179
  15 The Mating Game   195

Part Three The Human Connection

  16 Coatis Past   213
  17 Coatis Future   231

Acknowledgments   249

Appendix I. Common and scientific names of plants and animals   251

Index   255

Preface

The truck rolled to a stop under some large silver-leafed oaks. After a long day of driving, it was good to just stop moving. I rolled down my window and drank in the fragrance of the Huachuca Mountains, an earthy mix of juniper and humus, of life and death and rebirth. I closed my eyes and listened, trying to get the noise of the highway out of my head. I heard Garden Creek softly murmuring in the distance as a breeze whispered through the oak leaves. As I absorbed the gentle sounds of the mountains, another sound entered my awareness: a light patter of small objects landing on fallen leaves. It sounded like raindrops, or shells of nuts dropped by a feeding tree squirrel. It wasn't raining, so I opened my eyes and peered into the shaded forest to see if I could spot the squirrel.

But the sounds were not made by squirrels. I spotted several meter-long (3-foot) tails hanging from the tree branches and recognized them instantly. Up in the trees, just a dozen meters from my truck, a troop of coatis was feeding on Arizona madrone berries. I carefully opened the truck door and crept out to get a closer look. An alarm snort went out, and the forest erupted with sound as the coatis scrambled down tree trunks and crashed through branches and leaf litter to try to get as far from me as they could. I counted 16 as they fled and probably missed some. Before I returned to the truck, I realized I would have to greatly improve my stalking skills if I wanted to spend much time watching these animals. It was late November 1995, and after years of working out the funding and logistics, I had just moved to southeastern Arizona to begin a study on white-nosed coatis. Little did I know then that what started out as a curiosity about a tropical animal living in sub-tropical habitat would turn into an obsession that would occupy much of the next two decades of my life.

Why I conducted the study will be addressed in upcoming pages. I wrote this book to share my coati stories and observations, and compile information on coatis from other studies, most of which is locked away in theses and technical publications. My hope is that taking a look at all of the studies together will give a more holistic view of these fascinating animals and the environments they live in. While our attention is often drawn to the exotic, colorful, and charismatic species, the common and little species live equally interesting lives and have so much to teach us about this amazing planet we live on. None of these species live in isolation; all are part of an incredibly complex puzzle of life, dynamic in space and time, interwoven and interdependent. It's these relationships among the predators and competitors, the abiotic and symbiotic, that intrigue me. Perhaps this journey into the lives of these amazing and curious animals will help you share in my sense of wonder I feel with every trip into the field.

PART ONE
INTRODUCTION

1

What is a Coati?

The first sighting of a troop of coatis is not soon forgotten. Busy and garrulous, they move through the forest or scrub, poking their nose into every crevice, turning over every rock and log in their path. Brown and furry, they look like they were put together from spare parts: a monkey's tail, an anteater's nose, the claws of a bear. As William Beebe described the coati in 1921,[1] *"a long, ever-wriggling snout, sharp teeth, eyes that twinkle with humor, and clawed paws which are more skillful than many a fingered hand."* Groups of 15, 20, 30 or more adult females and youngsters fade in and out of the brush like shadows, staying connected with bird-like chirps or pig-like grunts. But what are these unusual animals?

Physical description

A coati is a streamlined version of a raccoon, with an extraordinary long nose and long tail. Currently, two kinds of coatis are recognized: lowland coatis and mountain coatis. Lowland coatis, which includes the white-nosed coatis of Central and North America and the brown-nosed coatis of South America, differ in the color of their snouts and the distinctiveness of the rings on their tails. Among white-nosed coatis, the muzzle is white and the rings on the tail are usually faint and hardly noticeable. Among brown-nosed coatis, the muzzle is dark brown or black and the rings are quite distinct, and some people refer to them as ring-tailed coatis.[2,3] They are also sometimes referred to as South American coatis, however, because there are at least three species of coatis in South America, I find "brown-nosed" to be the least confusing name. Mountain coatis, found in the Andes of northwestern South America, average smaller than lowland coatis in all dimensions. They tend to be more uniformly colored, without noticeable eye spots. Their tail is shorter and may or may not have distinct rings.[4]

Roughly the size of a large house cat, coatis come in various of shades of brown, from dark chocolate to strawberry blonde to a greyish tan. The coat color varies among individual coatis, and individuals change color throughout the year. The new coat, put on between summer and fall, is usually very dark, although it can appear brown to reddish in lighter individuals. As the coat ages and is exposed to sunlight and wear and tear, the color fades. This large variability in color led to the misguided designation of several coati species However, their color can vary within a locale, and even within a litter.[2]

White-nosed coatis usually have a dark mask punctuated by white spots around the eyes. The front edges of their small, round ears are also white. Some wear a bib of cream-colored hair on their throat and chest which may extend to the shoulders. Their lower legs and feet are black; the five toes on each foot bear sharp claws. The long tail is brown, darker at the tip, and has lighter rings. Brown-nosed coatis appear more variable in coat color than white-nosed coatis. In the southern part of their range, in southern Brazil and Argentina, they bear similar markings to white-nosed coatis. Here they are often found in two distinct color phases – a reddish-brown, and a greyish tan. In the northwestern part of their range, in Columbia and Ecuador, they are more uniform in color, often showing reduced eye spots. They resemble, in color, the mountain coatis. Where they overlap with mountain coatis, they can be difficult to tell apart.[5]

Adult male lowland coatis are larger than adult females. Male white-nosed coatis average about 5 kg (11 lbs) and adult females about 4 kg (9 lbs). Males are also slightly longer than females. White-nosed coatis average about 110-120 cm long (43-47 in), with their tails making up about 46% of the total length.[3,6] Brown-nosed coatis average just a bit smaller.[3] Mountain coatis are about half the size and show less sexual dimorphism than lowland coatis.[4] Although coatis vary little in size throughout their range,[7-10] coatis at higher altitudes and latitudes have longer fur and look larger. In addition, coatis that are frightened can erect their fur (piloerection), making them appear larger.

Coati weights change throughout the year, with males losing weight during the mating season, and females during lactation.[7] In addition, coatis can pack on the pounds when they have access to human food or high-calorie diets in captivity. I have heard several stories of 23-kg (50-pound) wild coatis, but these were estimates from afar, and no actual weights were obtained. Unlike raccoons, skunks, and bears, which put on a lot of weight during autumn, coatis cannot sleep for extended periods just living off their fat.[11] They need to eat daily, and this need for readily available food has major implications for where coatis live.

Coatis are members of the mammalian order Carnivora, Latin for "meat eater." But it is also the most diverse order of mammals and includes members that eat only leaves (giant panda), only fruit and nectar (kinkajou), as well as those who eat only meat (polar bears and most cats). Members of the Carnivora range in size from 25 gm (a few oz - least weasel) to more than 1,250 kg (2,700 lbs - walrus). They inhabit nearly every terrestrial biome, and seals, sea lions, walruses, and otters, all carnivores, inhabit oceans and rivers.

Awhile back, I was flipping through the television channels and hesitated for a moment on "Are you smarter than a 5th grader" just as host Jeff Foxworthy was asking a contestant, "A heron is a carnivore, true or false?" "False," I shouted at the TV, as I was thinking of the mammalian order Carnivora, and a heron is not even a mammal. But the contestant listened instead to his young advisors and said "True." And, indeed, he was correct. Herons eat fish and frogs and are therefore carnivores. Because of the confusion with meat eaters of other taxa, some biologists have begun using the term "carnivoran" to refer to members of the order Carnivora and, to avoid the rebuke of any 5th graders, I will follow their lead.[12,13]

Coatis belong to a guild of fruit and bug eaters, which also includes ringtails, raccoons, black bears, foxes, skunks, and many other carnivorans, plus opossums (a marsupial). Most of these include fruit and bugs (by which I mean arthropods, mollusks, and earthworms) in their diet in addition to other forms of meat and some vegetable matter, but coatis are fruit and bug specialists. Sure, they will feast on the occasional lizard or pile of carrion, but in almost every study of coati diet to date, fruits and bugs have made up 90% or more of their diet.

In addition to the characteristics that define most mammals (fur, the ability to bear live young, and produce milk), carnivorans are also defined by their teeth. Notably, this includes long, pointed canines and, among the true meat eaters, well-developed shearing molars and premolars known as carnassials. Cats are the most extreme meat eaters and have the most developed carnassials. Although dogs, such as coyotes and wolves, include some fruits and insects in their diets, their well-developed carnassials reflects a diet of mostly meat. The omnivorous carnivorans tend to have less well-developed carnassials, and fruit-eaters, such as the kinkajou, have completely flattened molars. Coatis and raccoons are in between, with reduced carnassials.

The skulls of male and female coatis differ in several ways. The male has a much more robust skull than the female; it is taller, wider, with a pronounced sagittal crest; the bony spine that runs along the midline of top of the skull. This boney crest allows for the attachment of strong muscles that close the jaw. Both sexes have an impressive set of canines – the upper canines are triangular and come to a sharp point, while the lower canines are long and saber-like. The backs of the lower canines rub against the fronts of the upper canines keeping the surfaces razor sharp. But the biggest difference between the skulls is the length of the lower canine. The lower canines of males are almost twice as long as those of the females.[6] These differences are not apparent in the molars and premolars, which are somewhat flattened and reflect the coatis omnivorous food habits. According to John Gittleman and Blaire Van Valkenburgh, the molars and premolars of carnivorans reflect the diet, whereas the canines reflect the social structure.[14,15] In this case, the larger lower canines of male coatis are indicative of a mating system involving intense male-male competition for mates. These lower canines then are not so much used for catching and killing prey, but rather as weapons used in battles to decide mating privileges. These differences in the skulls led one researcher to hypothesize that male coatis were more carnivorous than females, and perhaps reduced competition with the troops by having a different diet.[16] This has not been supported by any studies of coati diets.

Coati Society

Coatis are unique among the carnivorans, indeed among mammals, for their society. Or should I say societies. Coatis have two different social arrangements. Adult males are mostly solitary, whereas adult females group together with their offspring of the last couple of years. These groups, called "bands," "troops," "tribes," "clans," and assorted other names, can range in size from 4 to 50 (with rare sightings of more than

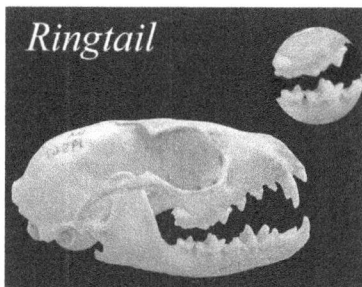

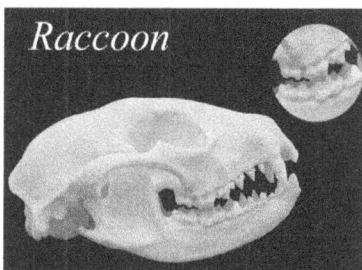

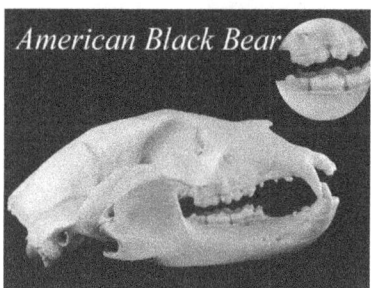

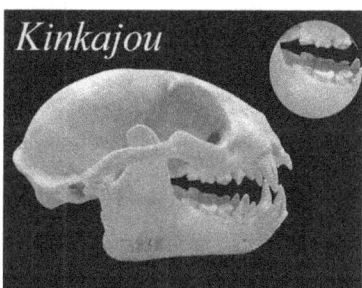

Skulls of various carnivorans reflect their diets. Note the shearing molars (carnassials) of the mountain lion and coyote (insets enlarged to show detail). The ringtail's carnassials are reduced, compared with mountain lion and coyote, and the black bear and raccoons are reduced even further. The kinkajou, which eats primarily fruit, has very flattened molars. Photos not to scale.

150!) but tend to average between 10 and 20 animals.[2,3,17] Coatis mature slowly compared to other animals their size, so it takes 2-3 years to achieve full growth and maturity to become an "adult." Males leave the troops when they become mature, which varies from 16 months to 3.5 years, but averages about 2 years.[17] Once males leave their natal troop, most take up a solitary existence, occasionally approaching and interacting with members of the troops, and aggressively attacking other males. Some studies, however, have reported males that appear to be members of troops.[18]

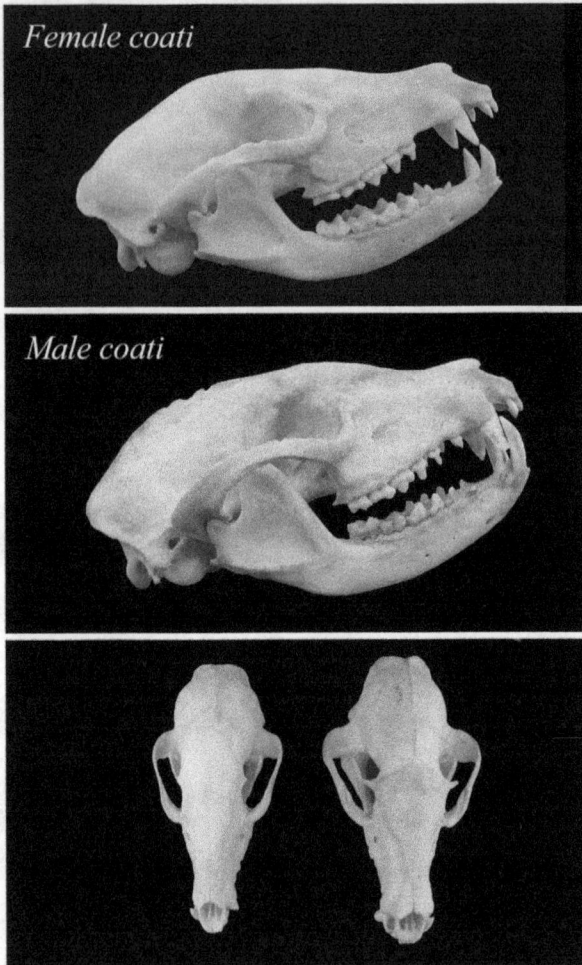

A comparison of the skulls of adult female (left) and adult male (right) white-nosed coatis from Arizona. Note the differences in the robustness of the skull and the length of the lower canine of the male coati.

What's in a Name?

Coatis go by many different names, which is not surprising given their distribution and the number of languages spoken by those who would name them. In addition, group-living coatis (females) are often given different names than solitary coatis (males). The term "coati" is a contraction of two words of Tupian Indian (Brazil) origin: "cua" meaning belt, and "tim" meaning nose. Although this has been reported to refer to the coatis' habit of sleeping with their nose on their belly,[2] coatis are no more inclined to sleep that way than any other animal. According to Virginia Holmgren, a better translation is, "*one with the flexible snout.*"[19] The Brazilian term "coati monde" (later coatimundi) refers to what was thought to be a different species of solitary coati. As this was later found not to be a separate species at all, the term really is a misnomer. As George Gaylord Simpson wrote in 1941, "*Coatimundi is also good Tupi, but the complication is unnecessary. The broad use of 'coati,' a shorter and more general term, seems preferable in English.*"[20] The term "coati" is preferred by scientists.

Regional names for white-nosed coatis include "gato en familia" and "gato solo" (Panama), "pizote" and "pizote solo" (Costa Rica, Honduras, Guatemala), and a variety of other names in various Mayan dialects. In central and southern Mexico, they are called "tejón" and "tejón solo," which, confusingly, is also the term for badger. In northern Mexico they are also called "chulos" and "cholugos."[21] In Arizona, I've heard them called chula bears, cootie monkeys, and coties. Some internet sources refer to them as "hog-nosed coons" and "snookum bears," but I have never heard those terms used. In South America, local names for brown-nosed coatis include "achuni," "tejón," "cuzumbo," "cochinigo," and "quati," and a large variety of names in local Native dialects.[3]

Regional names for mountain coatis include, "coatí de montaña," "guache," "cusumbo guache" and "guache de montaña," "guache de paramo," "guache de tierra fría," "zorro guache," "cusumbo" and "cusumbo de montaña," "runcho guache," "cusumbe" and "cusumbo mocoso," "cusumbo de paramo," and "cuchuche Andino."[22]

To the scientist they are known as *Nasua*, which is Latin for nose, and *Nasuella*, for little coati. Taxonomists originally designated males and females as separate species (*Nasua sociabilis* and *Nasua solitaris* – both referring to brown-nosed coatis).[2] The white-nosed coati (*Nasua narica*) of Central and North America is sometimes confused with the brown-nosed coati of South America (*Nasua nasua*). A scientist in the 1950's suggested that several species of coati in South America, which had been designated by coat color, were actually all the same species.[23] This was misinterpreted to mean that all coatis were the same species.[17] This has been very confusing for subsequent researchers and Mammalogists. In his book, *Mammals of Arizona*, Donald Hoffmeister uses both names in different parts of the book to refer to the white-nosed coati.[24] It is still not unusual to see coatis in

North America incorrectly referred to as *N. nasua*, but hopefully this will be corrected with time.

The Family Tree

The order Carnivora is typically divided into two suborders, the Feliformia (cat-like) and Caniformia (dog-like). The Feliformia includes the cats, civets, mongooses, and hyenas. Caniformia includes the families Canidae (dogs, foxes, wolves, jackals), Ursidae (bears and Giant Panda), Otariidae (eared seals), Odobenidae (walrus), and Phocidae (earless seals), Mustelidae (badgers, weasels, wolverines, honey badgers, tayras, grisons, ferrets), Mephitidae (skunks and stink badger), Ailuridae (red panda), and the Procyonidae (raccoons, coatis, ringtails, olingos, and kinkajou).

The first fossils of procyonids are from what is now Europe some 25 million years ago (mya), a time when tropical environments extended well into high latitudes. A diverse group of ancestral procyonids thrived in Europe and Asia until about 18 million years ago. The procyonids of Europe and Asia ultimately died out, but at least one group made it to North America and thrived, slowly moving south as northern latitudes began to cool and the tropics receded toward to equator.[27] The current species of procyonids in the Americas are descendants of that group. Ancestral procyonids appeared similar to modern coatis and olingos.[28]

North and South America split separately from the supercontinent Pangaea about 175 million years ago. The continents only reconnected in the last few million years, when the Panamanian land bridge appeared during times of relatively low sea levels. The rejoining of these two large continents resulted in a massive exchange of organisms, known as the Great American Biotic Interchange (or GABI), as plants and animals dispersed into new habitat. GABI has been a source of great fascination and controversy among scientists, supplying information about rates and directions of dispersal and evolution.

Coatis have been caught up in the middle of research on GABI, and recent research on coati genetics has indicated that the closure of the isthmus of Panama (which allowed organisms to travel overland between North and South America) occurred repeatedly and earlier than previously thought.[25,29,30] Traditional models suggested that the land bridge did not appear until 3-3.5 mya, with most mammals migrating south and colonizing South America 2.4-2.8 mya.[29,31] However, a new model, based on genetic evidence, suggests that the land bridge first appeared 23-25 mya, and

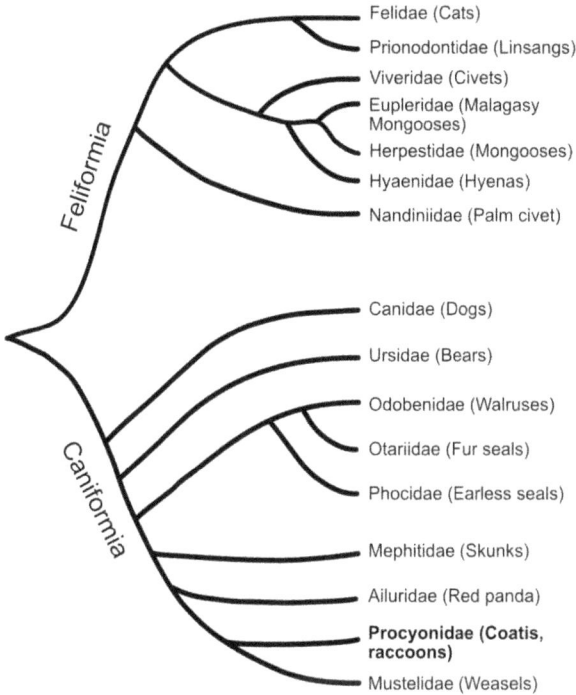

Phylogenetic tree:

- **Feliformia**
  - Felidae (Cats)
  - Prionodontidae (Linsangs)
  - Viveridae (Civets)
  - Eupleridae (Malagasy Mongooses)
  - Herpestidae (Mongooses)
  - Hyaenidae (Hyenas)
  - Nandiniidae (Palm civet)
- **Caniformia**
  - Canidae (Dogs)
  - Ursidae (Bears)
  - Odobenidae (Walruses)
  - Otariidae (Fur seals)
  - Phocidae (Earless seals)
  - Mephitidae (Skunks)
  - Ailuridae (Red panda)
  - **Procyonidae (Coatis, raccoons)**
  - Mustelidae (Weasels)

Phylogeny of the Carnivores, with Procyonidae highlighted.[25,26]

reached full closure 13-15 mya.[29] The most significant pulses of migration between continents, in both directions, occurred at 20 and 6 mya.[32]

Based on the latest genetic studies, the major divergence of coati species began in northern South America about six mya. During glacial expansion, many tropical areas dried out and became savannas, affecting animal dispersal and evolution.[31,33,34] Habitat changes caused by glaciations (including drying of tropical forests) and the upheaval of the Andean Cordillera probably caused the initial split between brown-nosed and mountain coatis, followed by the split between mountain coatis and white-nosed coatis.[30] From there, white-nosed coatis moved slowly north, with populations becoming further isolated by subsequent glaciations and mountain upheavals.[33] Brown-nosed coatis initially stayed east of the Andes, eventually inhabiting tropical forests all the way to Argentina. Brown-nosed coatis appear to have snuck through the lower passes of the northern Andes, and now occupy much of Columbia and Ecuador.[35] They overlap with mountain coatis at mid-elevations in parts of the northern Andes.[36] The genetics of coatis in Columbia and Ecuador are muddy and complicated, with close relationships among brown-nosed, white-nosed, and mountain coatis.[30,37] Columbia appears to be where most species of coatis arose, and continues to be a nexus of coati species, with at least three species and possibly more once more genetic research is completed. It should be noted that because procyonids are tropical forest dwellers, their fossils are scarce. Most remains decompose before they can fossilize. Therefore, there is much we don't know about the fossil history of the group, and phylogenies reconstructed with fossil histories are in conflict with the molecular data.[38,39]

The family Procyonidae, to which raccoons and coatis belong, includes some 14 species.[25] There are four species of olingos (belonging to the genus *Bassaricyon*), two species of ringtail (*Bassariscus*), two coatis (*Nasua*), two mountain coatis (*Nasuella*), one kinkajou (*Potos*), and three raccoons (*Procyon*). Previous research on morphology identified three groupings within the family: kinkajous and olingos were one related group; lowland coatis, mountain coatis, and raccoons were another group; and ringtails were a third group.[43] However, recent genetic analysis has revealed that is not the case at all. Coatis are most closely related to olingos (forming one branch of the genetic tree, or clade); raccoons and ringtails belong to a separate clade, and kinkajous form their own sister lineage.[38,41]

The similarities between raccoons and coatis and between olingos and kinkajous are examples of convergent evolution.[25,38] Olingos and kinkajous stayed in the trees and became frugivores. Raccoons and coatis took to the ground, becoming more omnivorous. Ringtails split the difference. The ringtail, *Bassariscus astutus*, is the most carnivorous of the family, feeding on small rodents and reptiles as well as fruit and bugs. It forages in trees and on the ground and is the only procyonid to move out of the forest; some ringtails live in rocky desert canyons. *Bassariscus sumacrasti*, commonly called the cacomistle, is more frugivorous and arboreal than the ringtail, although it looks similar. Both ringtails and cacomistles are nocturnal and not very social.

Coatis and raccoons show similar adaptations for a predominantly terrestrial lifestyle: their feet are better for digging than climbing, and they lack the forward-facing eyes that enhance stereoscopic vision for moving through the treetops. That said, both coatis and raccoons do spend a lot of time in trees, using them for feeding, resting, and giving birth.[41] Although superficially similar, coatis and raccoons differ in two important ways: coatis are diurnal (active during the day) and highly social, whereas raccoon are nocturnal (active at night) and less social.

Olingos and kinkajous are relatively small (only a couple of kg) and nocturnal. Although kinkajous are common throughout their range (northern Mexico through southern Brazil), they are seldom seen and not well-studied. They are the only procyonids to have a prehensile tail, and they also have an incredibly long tongue, used for foraging on nectar and fruits.[42-44] The four species of olingos are very poorly understood.[45] They range from Nicaragua to Peru and Brazil, and superficially resemble the kinkajou, but they are slightly smaller and lack the prehensile tail and long tongue.[41] They are also nocturnal, frugivorous, and they have an anal scent gland that produces defensive secretions that rival those of skunks.[19,46,47] In early 2013, a new species of olingo (the olinguito, *Bassaricyon neblina*) was identified in the Andes of Columbia and Ecuador.[45,48,49]

Species Distribution

As a group, coatis occur throughout the New World tropics and subtropics, wherever there is sufficient forest or scrub to provide cover and food. White-nosed coatis are found from southeastern Arizona and southwestern New Mexico to Panama, just barely crossing the border into Columbia.[50] On Cozumel Island, off the coast of the Mexico's Yucatan Peninsula, lives a smaller version of the white-nosed coati. It may have been brought to the island by the Mayans, some 2500 years ago.[51] Its status is uncertain; some researchers consider it a separate species, the Cozumel Island coati (*Nasua nelsoni*),[52,53] but genetic evidence suggests it is a subspecies of the white-nosed coati.[33]

The brown-nosed coati lives throughout the forested areas of South America, from Columbia and Venezuela to northern Argentina, below 2500 m (8300 feet) in altitude.[3,54] Only recently have detailed ecological studies been conducted on brown-nosed coatis, and most of those have focused on southeastern Brazil and along the Brazil-Argentina border.[55-60] Much more research is needed in other parts of its range. Recent genetic research indicates that there is substantial variation among populations throughout its range, suggesting there may be multiple species instead of just one.[30]

Currently assigned to a separate genus (*Nasuella*), mountain coatis live in the high Andes above 1300 m (4200 ft) in altitude. Recent genetic and

Distribution of coatis in North, Central and South America. Black dots show the locations of major coati studies.

morphological studies identified two species, the western mountain coati (*Nasuella olivacea*), found in Columbia, Ecuador, and Peru, and the eastern mountain coati (*Nasuella meridensis*), found only in Venezuela.[4] However, the latest genetic studies indicate that this may not be the case, rather that there are several distinct subspecies of *Nasuella olivacea* throughout the Andes, with none distinct enough to consider a separate species.[30] Mountain coatis are considered sister species to lowland coatis, with a lack of genetic support to keep them as a separate genus,[30] therefore they will probably be renamed to *Nasua* instead of *Nasuella*. Most of this book will focus on white-nosed and brown-nosed coatis.

Notes and References

1. Brand DD (1964) The coati or pisote (*Nasua narica*) in the archaeology and ethnology of Meso-America. *In*: Sobretiro del XXXV Internacional de Americanistas: Mexico. Mexico City, pp 193–202

2. Gompper ME (1995) Nasua narica. Mammalian Species 487:1–10

3. Gompper ME, Decker DM (1998) Nasua nasua. Mammalian Species 580:1–9

4. Helgen KM, Kays R, Helgen LE, Tsuchiya-Jerep MTN, Pinto CM, Koepfli K-P, Eizirik E, Maldanaldo JE (2009) Taxonomic boundaries and geographic distributions revealed by an integrative systematic overview of the mountain coatis, *Nasuella* (Carnivora: Procyonidae). Small Carnivore Conservation 41:65–74

5. Mirian Tsuchiya, personal communication.

6. Hass CC (1997) Ecology of white-nosed coatis in the Huachuca Mountains, Arizona, a preliminary study. Final report submitted to the Arizona Game & Fish Department, Phoenix, AZ. 1–52

7. Weights of white-nosed coatis from Panama: males, 5.1 kg (n = 51), females, 3.7 kg (n = 37); Jalisco: males 5.4 kg (n = 24), females, 4.2 kg (n = 42); Tamaulipas: males, 5.6 kg (n = 12), females, 3.4 kg (n = 4); Arizona: males 5.5 kg (n = 26), females 4.0 kg (n = 43).

8. Estrada A, Halffter G, Coates-Estrada R, Meritt DA Jr (1993) Dung beetle attracted to mammalian herbivore (*Alouatta palliata*) and omnivore (*Nasua narica*) dung in the tropical rain forest of Los Tuxtlas, Mexico. Journal of Tropical Ecology 9:45–54

9. Gompper ME (1996) Sociality and asociality in white-nosed coatis (*Nasua narica*): foraging costs and benefits. Behavioral ecology 7:254–263

10. Valenzuela D (1999) Efectos de la estacionalidad ambiental en la densidad, la conducta de agrupamiento y el tamaño del área de actividad del coatí (*Nasua narica*) en selvas tropicales caducifolias. Ph.D. Disertación, Instituto de Ecología, UNAM, México, D.F.

11. Mugaas JN, Seidensticker J, Mahlke-Johnson KP (1993) Metabolic adaptation to climate and distribution of the raccoon *Procyon lotor* and other Procyonidae. Smithsonian Contributions to Zoology 542:1–34

12. Gittleman JL (1989) Carnivore group living: comparative trends. *In*: Gittleman JL (ed) Carnivore behavior, ecology, and evolution. Cornell University Press, Ithaca, NY, pp 183–207

13. Hunt RU Jr (1996) Biogeography of the order Carnivora. *In*: Gittleman JL (ed) Carnivore behavior, ecology and evolution. Cornell University Press, Ithaca, NY, pp 485–541

14. Gittleman JL, Van Valkenburgh B (1997) Sexual dimorphism in the canines and skulls of carnivores: effects of size, phylogeny, and behavioural ecology. Journal of Zoology 242:97–117

15. Van Valkenburgh B (1989) Carnivore dental adaptations and diet: a study of trophic diversity within guilds. *In*: Gittleman JL (ed) Carnivore behavioral, ecology, and evolution. Cornell University Press, Ithaca, NY, pp 410–436

16. Smythe N (1970) The adaptive value of the social organization of the coati (*Nasua narica*). Journal of Mammalogy 51:818–820

17. Kaufmann JH (1962) Ecology and social behavior of the coati, *Nasua narica*, on Barro Colorado Island, Panama. University of California Publications in Zoology 60:95–222

18. Gompper ME, Krinsley JS (1992) Variation in social behavior of adult male coatis (*Nasua narica*) in Panama. Biotropica 24:216–219

19. Holmgren VC (1990) Raccoons in folklore, history and today's backyards. Capra Press, Santa Barbara, CA

20. Simpson GG (1941) Vernacular names of South American mammals. Journal of Mammalogy 22:1–17

21. Schoenhals LC (1988) A Spanish-English glossary of Mexican flora and fauna. Summer Institute of Linguistics, Hidalgo, Mexico

22. Balaguera-Reina SA, Cepeda A, Gonzalez-Maya JF (2009) The state of knowledge of western mountain coati *Nasuella olivacea* in Colombia, and extent of occurrence in the northern Andes. Small Carnivore Conservation 41:35–40

23. Cabrera A. (1957) Catalogo de los mamiferos de America del sur. Museo Argentino de Ciencias Naturales "Bernardo Rivadavia," Buenos Aires

24. Hoffmeister DF (1986) Mammals of Arizona. University of Arizona Press, Tucson, AZ

25. Koepfli K-P, Dragoo JW, Wang X (2017) The evolutionary history and molecular systematics of the Musteloidea. *In*: Macdonald DW, Newman C, Harrington LA (eds) Biology and conservation of the Musteloids. Oxford University Press, Oxford, U.K., pp 75–91

26. Zhou, Y, Wang, S-R, Ma, J-Z (2017) Comprehensive species set revealing the phylogeny and biogeography of Feliformia (Mammalia, Carnivora) based on mitochondrial DNA. PloS ONE 2017, 2(3) e0174902

27. Martin LD (1989) Fossil history of the terrestrial carnivora. *In*: Gittleman JL (ed) Carnivore behavior, ecology and evolution. Cornell University Press, Ithaca, NY, pp 536–568

28. Baskin JA (1989) Comments on New World tertiary Procyonidae (Mammalia: Carnivora). Journal of Vertebrate Paleontology 9:110–117

29. Nigenda-Morales SF, Gompper ME, Valenzuela-Galván D, et al (2019) Phylogeographic and diversification patterns of the white-nosed coati (*Nasua narica*): Evidence for south-to-north colonization of North America. Molecular Phylogenetics and Evolution 131:149–163

30. Ruiz-García M, Jaramillo MF, Cáceres-Martínez CH, Shostell JM (2020) The phylogeographic structure of the mountain coati (*Nasuella olivacea*; Procyonidae, Carnivora), and its phylogenetic relationships with other coati species (*Nasua nasua* and *Nasua narica*) as inferred by mitochondrial DNA. Mammalian Biology. https://doi.org/10.1007/s42991-020-00050-w

31. Bacon CD, Molnar P, Antonelli A, Crawford AJ, Montes C, Vallejo-Pareja MC (2016) Quaternary glaciation and the Great American Biotic Interchange. Geology 44:375–378

32. Bacon CD, Silvestro D, Jaramillo C, Smith BT, Chakrabarty P, Antonelli A (2015) Biological evidence supports an early and complex emergence of the Isthmus of Panama. PNAS 112:6110–6115

33. Nigenda-Morales SF (2016) Phenotypic and gene expression in the Virginia opossum (*Didelphis virginiana*) and phylogeography of the white-nosed coati (*Nasua narica*). Ph.D. Dissertation, University of California, Los Angeles

34. Woodburne MO (2010) The Great American Biotic Interchange: dispersals, tectonics, climate, sea level and holding pens. Journal of Mammalian Evolution 17:245–264

35. Guzman-Lenis AR (2004) Revision preliminar de la familia Procyonidae en Columbia. Acta Biologica Columbiana 9:69–76

36. Gonzalez-Maya JF, Vela-Vargas IM, -Alvarado JS, Hurtado-Moreno AP, Moreno C, Aconcha-Abril I, Zarrate-Charry DA (2015) First sympatric records of Coatis (*Nasuella olivacea* and *Nasua nasua*; Carnivora: Procyonidae) from Colombia. Small Carnivore Conservation 52 & 53:93–100

37. Ruiz-García M, Jaramillo MF, Shostell JM (2020) How many taxa are within the genus *Nasua* (including *Nasuella*; Procyonidae, Carnivora)? The mitochondrial reconstruction of the complex evolutionary history of the coatis throughout the Neotropics. Journal of Zoological Systematics and Evolutionary Research In press

38. Fulton TL, Strobeck C (2007) Novel phylogeny of the raccoon family (Procyonidae: Carnivora) based on nuclear and mitochondrial DNA evidence. Molecular Phylogenetics and Evolution 43:1171–1177

39. Forasiepi AM, Soibelzon LH, Gomez CS, Sánchez R, Quiroz LI, Jaramillo C, Sánchez-Villagra MR (2014) Carnivorans at the Great American Biotic Interchange: new discoveries from the northern neotropics. Naturwissenschaften 101:965–974

40. Decker DM, Wozencraft WC (1991) Phylogenetic analysis of recent procyonid genera. Journal of Mammalogy 72:42–55

41. Koepfli K-P, Gompper ME, Eizirik E, Ho C-C, Linden L, Maldonado JE, Wayne RK (2007) Phylogeny of the Procyonidae (Mammalia: Carnivora): Molecules, morphology and the Great American Interchange. Molecular Phylogenetics and Evolution 43:1076–1095

42. Kays RW (1999) Food preferences of kinkajous (*Potos flavus*): a frugivorous carnivore. Journal of Mammalogy 80:589–599

43. Kays RW, Gittleman JL (1995) Home range size and social behavior of kinkajous (*Potos flavus*) in the Republic of Panama. Biotropica 27:530

44. Brooks M, Kays RW (2017) Kinkajou: the tree-top specialist. *In*: Macdonald DW, Newman C, Harrington LA (eds) Biology and conservation of musteloids. Oxford University Press, Oxford, UK, pp 493–501

45. Helgen KM, Pinto CM, Kays R, Helgen LE, Tsuchiya MTN, Quinn A, Wilson DE, Maldonado JE (2013) Taxonomic revision of the olingos (*Bassaricyon*), with description of a new species, the Olinguito. ZooKeys 324:1–83

46. Kays R (2000) The behavior and ecology of olingos (*Bassaricyon gabbii*) and their competition with kinkajous (*Potos flavus*) in central Panama. Mammalia 64:1–10

47. Mendes Pontes AR, Rosas Ribeiro PF, Mendonca TM (2002) Olingos, *Bassaricyon beddardi* Pocock, 1921, in Brazilian Amazonia: status and recommendations. Small Carnivore Conservation 26:7–8

48. Cardona D, Meza-Joya FL, Colmenares J (2016) New olingo records (genus *Bassaricyon*) from the Colombian Andes. Small Carnivore Conservation 54:12–18

49. Gerstner BE, Kass JM, Kays R, Helgen KM, Anderson RP (2018) Revised distributional estimates for the recently discovered olinguito (*Bassaricyon neblina*), with comments on natural and taxonomic history. Journal of Mammalogy 99:321–332

50. Gonzalez-Maya JF, Rodriguez-Bolanos A, Pinto D, Jimenez-Ortega AM (2011) Recent confirmed records and distribution of the white-nosed coati *Nasua narica* in Columbia. Small Carnivore Conservation 45:26–30

51. Hamblin NL (1984) Animal use by the Cozumel Maya. University of Arizona Press, Tucson, AZ

52. McFadden KW, Gompper ME, Valenzuela DG, Morales JC (2008) Evolutionary history of the critically endangered Cozumel dwarf carnivores inferred from mitochondrial DNA analyses. Journal of Zoology 276:176–186

53. Cuarón AD, Martínez-Morales MA, Mcfadden KW, Valenzuela D, Gompper ME (2004) The status of dwarf carnivores on Cozumel Island, Mexico. Biodiversity and Conservation 13:317–331

54. Prieto-Torres DA, Herrera-Trujillo OL, Ferrer-Pérez A (2015) First record of *Nasua nasua* (Linnaeus, 1766) (Mammalia: Carnivora: Procyonidae) for the Zulia state, western Venezuela. Check List 11:1790

55. Aguiar LM, Moro-Rios RF, Silvestre T, Silva-Pereira JE, Bilski DR, Passos FC, Sekiama ML, Rocha VJ (2011) Diet of brown-nosed coatis and crab-eating raccoons from a mosaic landscape with exotic plantations in southern Brazil. Studies on Neotropical Fauna and Environment 46:153–161

56. Alves-Costa CP, da Fonseca GAB, Christofaro C (2004) Variation in the diet of the brown-nosed coati (*Nasua nasua*) in southeastern Brazil. Journal of Mammalogy 85:478–482

57. Beisegel B de M (2007) Foraging association between coatis (*Nasua nasua*) and birds of the Atlantic Forest, Brazil. Biotropica 39:283–285

58. Beisiegel B de M, Montovani W (2010) Habitat use, home range and foraging preferences of the coati *Nasua nasua* in a pluvial tropical Atlantic forest area. Journal of Zoology 269:77–87

59. Costa E, Mauro R, Silva J (2009) Group composition and activity patterns of brown-nosed coatis in savanna fragments, Mato Grosso do sul, Brazil. Brazilian Journal of Biology 69:985–991

60. Hirsch BT (2007) Within-group spatial position in ring-tailed coatis (*Nasua nasua*): balancing predation, feeding success, and social competition. Ph.D. Dissertation, Stony Brook University, Stony Brook, NY

2

Why Coatis? Why Arizona?

The idea for a coati study began high in the mountains of northern New Mexico. I was at my remote camp at almost 3600 m (12,000 ft) in the Pecos Wilderness, one of the study areas for my dissertation research on environmental factors affecting reproduction in bighorn sheep. Amy Fisher, who was then the bighorn sheep biologist for the New Mexico Game & Fish Department, packed in to visit for a few days during the summer of 1992. I had known Amy for a long time and knew that she had done research for her master's degree on bighorn sheep in southwestern New Mexico.

After long days spent traipsing around the alpine watching bighorn lambs, we would sit around the campfire and tell stories about our various adventures in the field. One of Amy's stories included looking for sheep from a perch high on a ridge in the Peloncillo Mountains and glancing down to see a troop of coatis foraging through the canyon bottom, tails waving as they moved out of sight. The topic quickly changed from bighorns to coatis because there are few areas where the two species occur together. Her story of encountering coatis in the Peloncillos stuck with me. Although I loved watching and studying bighorns, and loved being in their mountain habitats, I was ready for a new animal to study. One of my primary interests in biology is how the environment shapes social behavior and reproduction, especially the timing and duration of reproductive seasons.[1-3] Coatis are diurnal, which makes them more amenable to study than something only active at night. Earlier studies of coatis in the tropics revealed complex social behaviors rare among carnivorans, so they promised to be a fascinating subject. In addition, their presence in the United States meant I would not have to deal with the hassles of permits from foreign countries or the expenses of international travel.

I started making inquiries to biologists that I knew near the Mexican border, trying to find potential coati populations to study. Soon I received a phone call from Sheridan Stone, who worked in the Wildlife Management Office at Fort Huachuca, in southeastern Arizona. According to Sheridan, Fort Huachuca had a good population of coatis and potential funding was available from the Arizona Game & Fish Department. It took several attempts, but eventually we obtained the funding, and the study was ready to begin in earnest at the end of 1995. The study would focus on population dynamics, food habits, and reproduction of the coatis at Fort Huachuca.

The Huachucas

Coatis are, in general, a tropical animal. So, what aspects of their biology allowed them to exist in the arid mountains of southeastern Arizona? To answer that question, I needed to not only understand coatis, I needed to understand limitations placed upon them by climate, vegetation, and terrain.

The Huachuca Mountain range, home of Fort Huachuca, extends 40 km (25 mi) north from the Mexican border. Elevations rise almost 1,500 m (5,000 ft) from the valley floor to the highest point, Miller Peak, at 2,885 m (9,466 ft). The valley floors are a mix of grassland and desert scrub, while oak woodland blankets the slopes at elevations above 1,500 m (5,000 ft). A mix of oaks, pines and firs occur above 2,100 m (7,000 ft). Timberline at this latitude is higher than 3,650 (12,000 ft), so there is no alpine tundra in the Huachucas. Almost 1,000 species of plants, more than 80 species of mammals, 360 species of birds, and 71 species of reptiles are found in the Huachucas.[4] According to botanists Steve McLaughlin and Janice Bowers, 70% of the plants in the Huachucas have their origins further south in the Sierra Madre.[5,6]

The oak woodland is known as Madrean evergreen oak woodland, or more commonly by its Spanish name, encinal. It includes up to six species of oaks (*Quercus* spp.) whose ranges are all centered in the Sierra Madre or farther south. Unlike more temperate-zone oaks, the evergreen oaks keep their leaves through the winter, replacing them during the spring. They tolerate the occasional winter snow by having flexible branches. Hiking in the encinal after a fresh snowfall is a wet proposition: as the snow melts, the branches that were weighed down by the snow spring back into their normal position, launching "snow bombs." The encinal is an especially important habitat type for coatis at the northern extent of their range, supplying cover and food, and including important food trees such as Arizona madrone and alligator juniper, in addition to the oaks.

The Huachucas lie in an area known as the Madrean Archipelago. This is a series of north-south oriented mountain ranges that straddle the U.S.-Mexican border. The archipelago separates the Sierra Madre to the south and the Rocky Mountains to the north. It also divides the Sonoran Desert to the west from the Chihuahuan Desert to the east. Plants and animals from all four regions can be found here, making it one of the most biologically diverse areas of the United States. The archipelago consists of two dozen isolated mountain ranges. Each mountain range differs in size and altitude, and as a result, each differs a little in the composition of the flora and fauna. The more northern ranges tend to have more Rocky Mountain species, and more southern ranges have more Madrean species. Likewise, there is an east-west gradient, with more species from the Sonoran Desert on the west side, and more from the Chihuahuan Desert on the east.[6]

Geologically, the Huachucas and other ranges of the Madrean Archipelago are part of the Basin and Range province, a buckling of the earth's surface caused by the collision of the Farallon and North American plates.[4,7] Fault lines that run through the Huachucas have cracked, lifted and rearranged the layers of Precambrian granites and Cambrian limestones, and intermixed them with assorted other volcanics.[4] Surface erosion has not only moved much of the looser material into the valleys, but has also increased the porosity of some of the limestone layers – opening fissures and caves that coatis, bats, and other wildlife use for shelter, and also affecting the way water moves through the mountains.

There are pronounced altitudinal effects, too. The mountains are high enough to have large differences in temperatures between the valley floors and the mountain tops. Adiabatic cooling results in temperatures that may be up to 7ºC (20ºF) cooler at the crest than at the base of the mountains.[8,9] This cooling air results in more moisture; the crest of the Huachucas receives an average of 53 cm (21 in) of precipitation each year, 15 cm (6 in) more than the valley floor.[10] Fifty to 60 percent of the annual

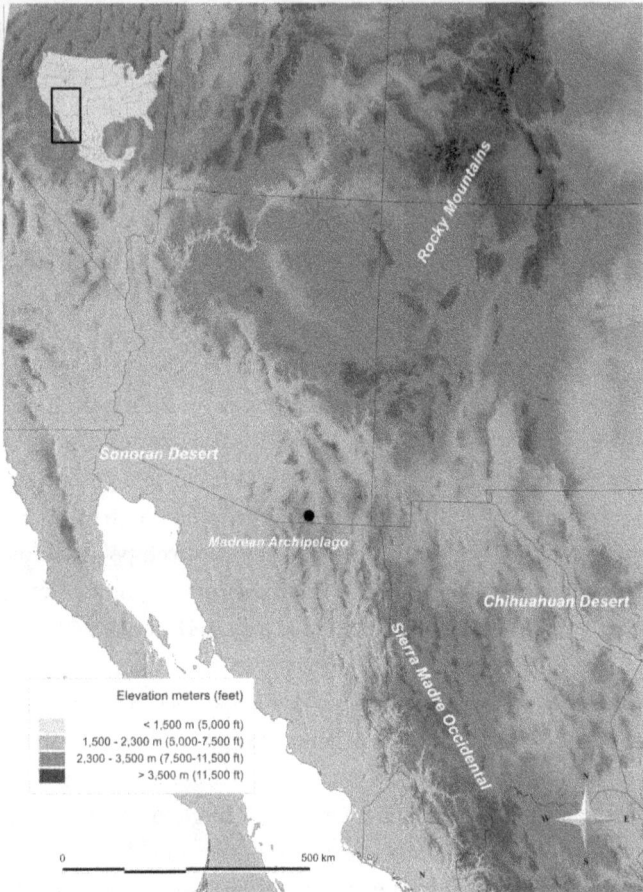

Location of the Huachuca Mountains (black dot), and topography of the Madrean Archipelago region.

rainfall occurs during July through September. This period, known as the North American Monsoon, results from moisture coming northward from the Gulf of Mexico and the Gulf of California, and usually arrives in the form of afternoon thunderstorms. Forming over the mountaintops nearly every afternoon, these storms can result in tremendous downpours and flash floods. This warm, wet weather is a boon to plant growth and everything that depends on plants.

Nearly everything in these southern mountains revolves around the monsoon. Many birds and most of the mammals time reproduction to co-incide with the flush of available food following the onset of the rains. Hills that were dry and brown since October become lush and green within a couple of weeks. Dry, crackly underbrush is replaced with lush green grass, forbs, and vines. Even mule deer and white-tailed deer fawns are not born until July or August, much later than their northern counterparts. Insects appear in swarms, supplying an important source of protein for baby birds and mammals. Coatis, too, are creatures of the monsoon.

The biological make-up of each mountain range in the archipelago is largely determined by latitude, longitude, and altitudinal extent. But to really understand the diversity of the Madrean Archipelago and its dynamic nature requires an imaginary trip back in time, to the Pleistocene. Although the continental ice sheets never came farther south than the northern Great Plains and Ohio River valley, mountain glaciers extended through the Sierra Nevada, Rocky Mountains, and even into southern New Mexico. The chilly breath of the glaciers was felt well into northern Mexico.[11] Beginning some 1.8 million years ago, the Pleistocene began like a slow glacial tide, extending its reach during pluvial periods, and waning during interglacials. The Madrean Archipelago was the shore to this ancient tide – not the glacial ice, but its cold breath. During glacial periods, cool, wet winter conditions allowed the survival of the Rocky Mountain pine forests much lower in elevation and much farther south. What is now grassland was once forest and woodland; desert basins were once lakes and marshes. Summers were cooler and drier, with fewer thunderstorms, and during some periods, overall precipitation was less than today.[11] The encinal, with its need for warm summer rains, retreated back into the Sierra Madre. During the interglacials, conditions were reversed. The cooler weather retreated, the summer monsoons extended farther north again, and the encinal returned. Remnant Rocky Mountain forests survived only on high mountain tops.

This process repeated itself ten to fifteen or more times. Each wave differing slightly in its extent and duration, as the flora and fauna of the Rockies landed on shores of the Sierra Madre. Glacial periods lasted some 30,000 to 100,000 years, while the interglacials were generally shorter, lasting 10,000-30,000 years.[12] Interestingly enough, according to Tom Van Devender, even though the average temperature was much cooler during the glacial periods, cold temperatures were less extreme. The glaciers blocked the frigid air from the Canadian Arctic that now periodically makes its way down the east side of the Rockies, bringing killing frosts well into northern Mexico. So, while the average temperature was much colder, the minimum temperatures did not get as low as they get today.

These slightly warmer minimum temperatures allowed some of the subtropical vegetation that established during the interglacials to survive some of the glacial periods,[12] resulting in some relict populations of thorn scrub much farther north than might be expected. As a result, some glacial flotsam endures at the highest altitudes and cool, sheltered drainages, while subtropical thorn scrub jetsam lingers in warm humid pockets. Animals, too, responded to these changes in the habitat: as recent as 100 years ago, both grizzlies and jaguars roamed the Madrean Archipelago.[13]

The tropics also felt the impact of the Pleistocene, as temperatures cooled a few degrees, and more importantly, rainfall decreased significantly. Areas of low-lying tropical forest became grassy savannas or scrub habitats, forcing many forest dwellers (and the forests themselves) into "islands" of wetter habitats.[14,15] In other words, what appears now to be continuous tropical forest was a mosaic of forest, grassland, and scrub during glacial periods. Even in these tropical habitats, landscapes were always changing, and the amount of available coati habitat came and went with each glaciation.[16]

The last ice age began its retreat about 15,000 years ago. It took several thousand more years for the present climate pattern to establish, and it could be reasonably argued that change is still in process, as the cold wintery climate of the glacial age gives way to a stronger summer precipitation pattern. This summer monsoonal pattern currently stops near the Mogollon rim, a basaltic escarpment that separates the Madrean Archipelago from the Colorado Plateau and southern Rockies. Although the southern Rockies and some of the Great Plains receive significant summer thunderstorms, most of their annual precipitation occurs during winter and spring. This dividing line between the summer precipitation and winter precipitation dominance defines the northward extent of a number of tropical and subtropical species, not only the coatis, but also hooded skunks[17] and dozens of species of birds.

As the glaciers began their retreat, southeastern Arizona looked much like central Wyoming does today except that the fauna was quite different: mastodons, dire wolves, ground sloths, and other giant mammals roamed the area. Humans appeared with the withdrawal of the glaciers and shortly thereafter the giant Pleistocene beasts began to disappear. Remains of early hunter-gathers, the Clovis people, have been found associated with mammoth kill sites along the San Pedro River, just east of the

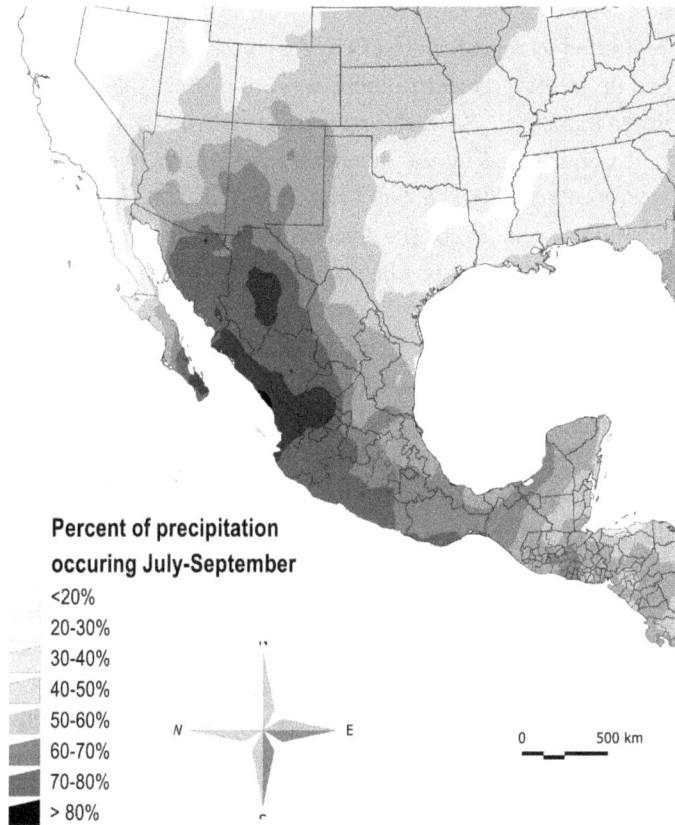

Percent of precipitation occuring July-September
<20%
20-30%
30-40%
40-50%
50-60%
60-70%
70-80%
> 80%

N — E
0   500 km

The percent of annual precipitation that falls during July through September in Mexico and the southwestern U.S.[18]

Huachucas, dating back some 13,000 years.[19,20] Other archaeological sites reveal an almost continuous occupation of the area ever since, not only by hunter-gathers, but by early farmers such as the Hohokam and Sobaipuri.[21] These early landscape architects developed irrigation systems, used fire to alter forest and grassland, established plantings of agaves and other food plants, and impacted wildlife populations through localized hunting as well as indirect impacts on the landscape.[21,22]

The Spanish explorer Francisco Vásquez de Coronado took his first steps into what is now the United States near the base of the Huachucas in 1540, arriving just as the Sobaipuri farmers were beginning to feel pressure from the Apaches, newly arrived from the north. Ultimately the Apaches took control of the area and kept Spanish settlers at bay for almost three hundred years. Spanish and Mexican occupancy during the

1700 and 1800's brought ranching of cattle, goats, and horses to the area, but were limited by repeated attacks by Apaches.[23] The ending of the Civil War freed up military power to move west to protect the growing cattle and railroad industries. Several forts were established in the area, including Fort Huachuca, and by the late 1800's the Apache were sequestered onto reservations in north-central Arizona and New Mexico.[24]

Following the removal of the Apache and the completion of the railroad to Tucson in 1880, settlers began to pour into the area in large numbers. The settlers brought cattle and sheep. Hundreds of thousands of them. Exact numbers are difficult to obtain, but there is widespread agreement that there were far too many for an area prone to periodic drought.[23] In many places, the land still has not recovered.[24,25] Wildfire suppression, mining, fuel wood cutting, and housing developments have also had their impacts upon the landscape.[23-25]

Amidst this ever-changing landscape appears the coati – a tropical animal in a non-tropical habitat. It was first recorded by settlers in 1892,[26] leading to the suggestion that they responded to the drastic habitat changes that occurred around that time.[25] They left their tracks in the encinal, pine forests, and riparian forests that snaked through the canyon bottoms. They eventually appeared in the canyons of the Mogollon rim, some 145 miles north of the Mexican border.[27] But their tracks are filled with mysteries: what does a coati eat, in a land with no papaya? Why do they seem to occupy an area for months at a time, only to fade like shadows into the forest? Are they recent immigrants, moving north as the climate warms?

My study was centered at Fort Huachuca (then formally known as the United States Army Intelligence Center & Fort Huachuca, or in military-acronymic, USAIC&FH), but when the coatis refused to recognize political boundaries it spilled onto the adjacent Coronado National Forest and The Nature Conservancy's Ramsey Canyon Preserve. Fort Huachuca, a military reservation of some 288 km² (71,000 ac), occupies the northeastern quarter of the Huachuca Mountains and some of the adjacent valley to the east, all the way to the San Pedro River. Fort Huachuca turned out to be a particularly good study site, due to its size, long history of ecological research, and supportive staff. In addition, during World War II many dirt roads had been carved into the mountainsides to act as fire breaks. While visually unappealing and ecologically damaging, they also supplied good foot access to rugged canyons and hillsides in the mountainous part of the Fort.

This made it much easier to find and monitor the coatis and their predators.

Doing research on a military reservation had its advantages and disadvantages, beyond those normally experienced in field work. There were permits to obtain and extra paperwork. Field work took a backseat to military training exercises, which usually did not affect my work in the canyons, but occasionally areas were off-limits when training was in progress or live-fire exercises were being conducted near the mountains. I was also stopped by Military Police repeatedly and asked if I was transporting drugs. The Fort was also enclosed by a barbed-wire fence, and access was through guard stations at the main entrances. This meant access was much more restricted than in the surrounding national forest. Usually this was not a problem, although access was extremely limited in the weeks following the terrorist attacks in New York City in 2001.

The mid to late 1990's were also a time of rapid increases in the smuggling of drugs and people across the border from Mexico into the United States. It was common to run into immigrants and smugglers throughout southern Arizona, even as the number of Border Patrol agents was rapidly increasing. Most of my biologist friends at some point in their field work ran across bodies of unfortunate souls who perished on the trek north. I did not find any deceased immigrants, but I frequently found sign of live ones, and occasionally the sign-makers, and even once encountered some scouts for one of the drug cartels, in my travels in and around Fort Huachuca. Law enforcement agencies generally recommended that biologists working near the border be armed and not to work alone. I followed neither recommendation, but just remained vigilant of my surroundings. During most of my coati study, Fort Huachuca felt like one of the safest places to be. After all, what smuggler would want to risk capture by crossing a well-guarded military reservation? But by the late 1990's, it appeared as though the military and Border Patrol were having some territorial disputes in terms of interdiction. No one was patrolling the Fort looking for smugglers or immigrants, and it did not take them long to figure out that the Fort was pretty much a free zone. By the early 2000s, the results of this strategy were clear – large piles of trash, clothing, empty water jugs littered every canyon and even the tops of the peaks. It was heartbreaking to see what amounted to a wilderness area spoiled this way.

So, despite, or perhaps because of, the constant changes in the environments of southern Arizona and northern Mexico, coatis persist. Their populations rise and fall, for a variety of reasons, but they contribute to a highly diverse fauna that we are only beginning to understand.

Notes and References

1. Hass CC (1993) Reproductive ecology of bighorn sheep in alpine and desert environments. Ph.D. Dissertation, University of North Dakota, Grand Forks

2. Hass CC (1995) Gestation periods and birth weights of desert bighorn sheep in relation to other Caprinae. The Southwestern Naturalist 40:139–147

3. Hass CC (1997) Seasonality of births in bighorn sheep. Journal of Mammalogy 78:1251–1260

4. Shaw HG (1999) Garden Canyon watershed. A vision and a mission. The Juniper Institute, Chino Valley, AZ

5. Bowers JE, McLaughlin SP (1994) Flora of the Huachuca Mountains, Cochise County, Arizona. *In:* Biodiversity and management of the Madrean Archipelago. USDA Forest Service Proceedings RM-GTR-264., pp 136–143

6. McLaughlin SP (1994) An overview of the flora of the Sky Islands, southeastern Arizona: diversity, affinities, and insularity. *In:* Biodiversity and management of the Madrean Archipelago. USDA Forest Service Proceedings RM-GTR-264, pp 60–70

7. Coblentz D, Riitters K (2005) A quantitative topographic analysis of the Sky Islands: a closer examination of the topography-biodiversity relationship in the Madrean Archipelago. *In:* Biodiversity and management of the Madrean Archipelago II. USDA Forest Service Proceedings RMRS-P-36, pp 69–74

8. Jerome (1978) describes a cooling rate of 2-5°F per 1000 ft of elevation gain, depending on whether moisture is condensing out of the air or not.

9. Jerome J (1978) On Mountains. Harcourt Brace Jovanovich, New York, NY

10. 14.7 in of rain at the Central Meteorological Office near the Sierra Vista Airport (37.3 cm, elev. 4667 ft.; data from 1974-1999), and 20.8 in (52.8 cm) at a site in upper Garden Canyon at 6700 ft; data from 1988 to 1999), data provided by the Army Electronic Proving Ground.

11. Van Devender (1990) Late quaternary vegetation and climate of the Chihuahuan Desert, United States and Mexico. *In:* Betancourt JL, Devender TRV, Martin PS (eds) Packrat middens: the last 40,000 years of biotic change. University of Arizona Press, Tucson, AZ, pp 104–133

12. Van Devender TR (2000) The deep history of the Sonoran Desert. *In:* Phillips SJ, Comus PW (eds) A natural history of the Sonoran Desert. Arizona-Sonora Desert Museum, Tucson, AZ, pp 61–69

13. Hoffmeister DF (1986) Mammals of Arizona. University of Arizona Press, Tucson, AZ

14. Ceballos G, Arroyo-Cabrales J, Ponce E (2010) Effects of Pleistocene environmental changes on the distribution and community structure of the mammalian fauna of Mexico. Quaternary Research 73:464–473

15. Piperno D R, Jones JG (2003) Paleoecological and archaeological implications of a late Pleistocene/early Holocene record of vegetation and climate from the Pacific coastal plain of Panama. Quaternary Research 59:79–87

16. Nigenda-Morales SF (2016) Phenotypic and gene expression in the Virginia opossum (*Didelphis virginiana*) and phylogeography of the white-nosed coati (*Nasua narica*). Ph.D. Dissertation, University of California, Los Angeles

17. Hass CC, Dragoo JW (2017) Competition and coexistence in sympatric skunks. *In*: MacDonald DW, Newman C, Harrington LA (eds) Biology and conservation of Musteloids. Oxford University Press, Oxford, UK, pp 464–477

18. Data from the Global Historic Climate Network (http://www.ncdc.noaa.gov/oa/ncdc.gov; accessed 10/2010).

19. Martin PS, Greene HW (2005) Twilight of the mammoths: ice age extinctions and the rewilding of America. University of California Press, Berkeley

20. Lucas SG, Morgan GS (2005) Pleistocene mammals of Arizona: an overview. *In*: Heckert AB, Lucas SG (eds) Vertebrate paleontology in Arizona. Albuquerque, NM, pp 152–157

21. Spoerl PM, Ravesloot JC (1994) From Casas Grandes to Casa Grande: prehistoric human impacts in the Sky Islands of southern Arizona and northwestern Mexico. *In*: Biodiversity and management of the Madrean Archipelago. USDA Forest Service Proceedings RM-GTR-264., pp 492–501

22. Mann CC (2005) 1491: new revelations of the Americas before Columbus. Vintage Press, New York, NY

23. Bahre CJ (1991) A legacy of change: historic human impact on vegetation of the Arizona borderlands. University of Arizona Press, Tucson, AZ

24. Wilson JP (1995) Islands in the desert: a history of the uplands of southeastern Arizona. University of New Mexico Press, Albuquerque, NM

25. Gehlbach FR (1993) Mountain islands and desert seas: a natural history of the U.S.-Mexican borderlands. Texas A&M University Press, College Station, TX

26. Kaufmann JH, Lanning DV, Poole S (1976) Current status and distribution of the coati in the United States. Journal of Mammalogy 57:621–637

27. Frey J, Lewis J, Guy R, Stuart J (2013) Use of anecdotal occurrence data in species distribution models: an example based on the white-nosed coati (*Nasua narica*) in the American Southwest. Animals 3:327–348

3

How to Catch a Coati

Although coatis are diurnal, they spend most of their time in thick forest or brush where they are difficult to see. In addition, coatis in Arizona cover huge areas. To find and watch them regularly, I needed to attach radio collars to individuals. But first, I had to capture them. Other researchers have used food to habituate the coatis, so they could approach the animals and observe them. Coatis in Arizona are hunted and, although the hunt was suspended at Fort Huachuca from 1996 to 2001, I was afraid that teaching coatis to approach people for food would make them vulnerable to getting shot. The coatis at Ramsey Canyon Preserve were somewhat habituated to people, and at times you could get quite close, but the other troops and males were very shy and usually fled when they detected human presence.

It took several months of trapping before the first coatis walked into the traps. Until we had animals with radio collars to follow, I and my assistant, Jason Roback, familiarized ourselves with the study area by exploring every backroad and trail, learning the lay of the land and looking for coati sign. We checked the live traps as soon as we arrived in the morning, and right before we left at the end of the day. We caught ringtails, opossums, and spotted and hooded skunks, but not coatis. We found their tiller-like diggings and occasionally got a brief glimpse of a troop (sometimes frustratingly close to the traps), but no captures. Finally, after weeks of disappointingly empty traps, trying many different trapping sites

31

and baits, suddenly our luck changed. One afternoon in early March 1996, after Jason and I had both spent a long day hiking and searching for signs of coatis, we found seven coatis in eight traps (one trap sprung but empty). It took us several hours to anesthetize,[1] measure and mark the two juveniles and five adult females, and we didn't finish until long after dark. But it finally meant that the study was underway in earnest. A couple of days later, we caught four more adult females at the same site. We would later learn that this was not unusual – coatis would often disappear from portions for their range for months at a time, making trapping very unpredictable. I was able to follow a couple of those animals captured and marked in March of 1996 until June of 2000.

Most of my earlier research had been conducted without capturing and handling animals, and I prefer it that way. Capturing animals is stressful; there are risks of injuries to the animals and the researcher. Trapping, anesthetizing, and attaching radio collars can influence behavior.[2] But reading the reports of previous researchers and knowing from previous trips to the Huachucas how difficult it would be to find the coatis on a regular basis, I decided that radio collars would be necessary. Capturing the animals provided added benefits by allowing me to collect information on body size and overall health. I could also collect blood samples to be sent to labs for analysis of exposure to rabies and distemper and for hormone analysis. Other small-to-medium-sized carnivorans captured in the traps supplied added data on distribution, body size, and health of the carnivoran community. The basic biology of some of these animals, such as hooded skunks, was relatively unknown. I was able to keep the coati study going by obtaining additional funding to work on skunks.

Capturing coatis

I used raccoon-sized wire mesh traps placed along canyon bottoms, usually near the dirt roads that supplied access to the canyons. I tried a variety of baits, including fruits and marshmallows. Jerry Pratt, who was the first wildlife manager at Fort Huachuca back in the 1950s and who had raised a few orphaned coatis, told me that they went crazy for cantaloupe. But cantaloupe as bait attracted more ants than coatis. David Valenzuela, who was studying coatis on the coast of Jalisco, said he used sardines in tomato sauce. Not only did that not work on my coatis, but I also watched a female coati carefully enter a trap, sniff the sardines, and back out of the trap. Perhaps fish is so foreign to Arizona coatis that they do not recognize it as food.

After much trial-and-error, I ended up using peanut butter and dried cat food for most of the year, and onion bagels and apples during the monsoon. The onion bagel suggestion came from Jason—one day when he sat down to eat a lunch of an onion bagel, a male coati that had become accustomed to handouts tried to steal it from him. An intense tug-of-war ensued, but ultimately they shared the bagel. For the most part, the coatis were not very picky, and I think they sometimes entered the traps more out of curiosity rather than attraction to the baits. Indeed, one morning when I was checking traps, I came across a ringtail in one of the traps. As I was getting ready to anesthetize it, I heard coatis scrambling down the cliff above me. I chased the ringtail out of the trap but did not have time to retrieve the bait from my truck, so I set the trap without any bait. I retreated, and sure enough, a few moments later a couple of coatis had entered the traps, including the trap with no bait in it.

Handling and radio-collaring

A coati in a trap takes on the demeanor of a pissed-off badger. Once I determined that it was an animal that I had not handled recently, I used a flat board as a plunger to gently push the animal to one end of the trap.[3] This helped restrict the animals' movements and gave me better access to a flank for the injection, which was done with a hand-held syringe though the wires of the trap. If all went well, the animal was unconscious within a few minutes and ready for handling. I tried to use only enough drug to keep the coati under for about 25 minutes. This was just enough time to do all the procedures and get the coati back in the trap to recover. If the coati had been handled recently, I simply opened the door of the trap, making sure to keep my fingers away from sharp teeth and claws, and the coati bounded away.

After I pulled the anesthetized animal out of the trap, I checked its overall condition. I determined its sex, which is quite easy for a coati in the hand, much more difficult when they are dodging through the brush. I measured their total length, the lengths of their tail, ear, hind foot, and lower and upper canines. I combed through their fur and counted the numbers of ticks, fleas, and mites. I carefully poked a needle into their jugular vein and withdrew a few cc of blood. I attached one or more eartags, which bore the coatis' identification numbers. Many of the adult animals received radio collars which were wrapped with colored plastic tape, some received colored dog collars, and some colored eartags. The marks allowed me to identify individuals through binoculars when I could get close. None of these marks were ideal; the tape on the radio collars wore out, the dog collars faded, and coatis were quite adept at ripping out eartags.

In addition to Fort Huachuca, I trapped and radio-collared coatis at The Nature Conservancy's Ramsey Canyon Preserve, just south of the Fort Huachuca boundary. I also tried trapping on the San Pedro Riparian National Conservation Area but was only successful in capturing skunks. Between October 1994, when I started collecting preliminary data and September 2002 when my final grant ended, I handled 295 carnivorans and opossums in and near the Huachucas and along the San Pedro River. In all, not counting recaptures, I handled 76 coatis (45 females, 31 males), 65 hooded skunks, 38 Sonoran opossums, 38 striped skunks, 35 ringtails, 18 spotted skunks, 11 gray foxes, 10 raccoons, 3 hog-nosed skunks, and 1 bobcat. I also captured gray squirrels, rock squirrels, pack rats, cottontails, and a couple of house cats, but these were all released without handling.

Besides the risk of being bitten or scratched, trapping had other risks. Once, I released a rock squirrel from a trap, and went back to my truck to get some more bait. When I returned to the trap, a black-tailed rattlesnake (one of at least eight species of rattlesnakes in the area) had taken up residence in front of the trap, probably drawn by the scent of the squirrel. It showed no inclination to move, so I waited until later in the day to rebait the trap. Exposure to poison ivy, chiggers, and scorpions also posed challenges.

Coatis varied widely in their response to traps. Many animals I caught repeatedly, and a couple of males entered the traps daily until they finally got tired of the routine. I was lucky enough to be able to catch a few animals just often enough to replace their radio collars when the batteries were due to expire (18-26 months) and was able to keep track of these individuals for up to five years. A couple of animals really seemed turned off by the experience, and no matter what I did, what bait I used or where I placed the trap, they would not re-enter a trap. I even saw one particularly shy female pushing other troop members away from the traps.

Some coatis figured out how to get the bait out of the traps without getting caught – by rolling the trap over onto its top. That way gravity pulled the door in the wrong direction, so the coati could safely go in and take the bait. Trying to foil the coati's attempts to foil my trapping was a never-ending game. One of the most effective tricks was surrounding the trap with large rocks to make the trap difficult to roll. If I caught a coati, ringtail, or opossum, it was easy to pick up the trap by the handle and move it to someplace where I could manage the rest of the handling. But catching a skunk was another matter – picking up the trap was a sure way to get sprayed.

Most of the skunks were soundly sleeping by the time I checked the traps in the morning. But any disturbance to the trap woke them up. Although the first few skunks I handled were nerve-wracking, I quickly learned that skunks were much more afraid of me than I was of them. If I moved low to the ground, and very slowly, they would usually settle down. I used a jab stick (a syringe at the end of a pole), carefully threaded through the mesh of the trap, to anesthetize the skunks. Once I approached the skunks and let them settle down, I could inject them and continue with handling once they were asleep. But some skunks were not keen on the idea and dodged the needle or bit at it. Skunks were the most challenging of the animals to handle, but in the end, I handled skunks more than 200 times, and was never sprayed. Skunks taught me a lot about approaching animals with respect and compassion.

I tried other methods of capturing coatis, but the results were not particularly good. A couple of males were tame enough that they could be approached close enough to use a jab stick. But a coatis' hide is so tough that it was difficult to get enough drug into them before they jumped away. With one male (M14), I was able to get sufficient drug injected that he became sleepy, so I could approach and administer another shot to knock him out enough for handling. The other male (M37) received enough to make him a little dopey, but when I slowly approached him again, he roused himself enough to scamper out of range. I chased him around for a while, then gave up and retreated down the hill, staying close to keep an eye on him. I was a little worried that his impaired condition might make him more vulnerable to predators, and I hoped my presence would deter them. However, what happened next was not what I expected.

As M37 sat on the hill above me, sleepily watching me watching him, a streak of brown came out of nowhere and bowled him over in an eruption of squeals, growls, screams, fur, and blood. Another male coati had attacked M37 with no warning, no preliminaries. They tumbled down the hillside, squealing, biting, and scratching, then as quick as it began, they split, with M37 going down canyon and the other male (unmarked) heading back the way he came. The whole encounter took less than a minute. M37 trotted off down the canyon, with no more sign of any dopiness – I think the adrenaline caused by the fight overrode the effects of the drug. He added a few more cuts and scrapes to his collection of scars, but otherwise was ok. But it let me know that I really needed to protect the coatis while they were recovering from anesthesia.

Based on relocation data, the coatis belonged to nine troops. I tried to place two or three radio collars in each troop, but that was not always possible, and at times the coatis were being killed by predators or the radio collars were failing as fast as I could attach them.[4] The battery life on the collars was 16-28 months, but some ceased to work before that. If I already had three functional collars in a group (and knew which group members I had in the traps), I would attach only a colored dog collar or colored eartags to additional animals. The troops had overlapping home ranges, and some areas were used by more than one troop. Some parts of Garden Canyon, for example, were used by four different troops at different times, so if I did not see the coatis near the traps, or a marked animal did not get caught in one of the traps, I could not tell which troop an animal belonged to until it was collared and rejoined its troop.

I used small VHF radio collars on the coatis. These were custom made for coatis with no external antenna for the coatis to chew off as they groomed each other.[5] The collars acted like their own little radio stations, each broadcasting on a unique frequency. Using a portable antenna and a radio receiver specially designed to pick up those frequencies, I could detect signals from the radio collars up to a mile away. The high frequencies broadcast by the collars were attenuated by rocks and vegetation and worked best in line of sight. The signals also bounced off solid rock, and the limestone cliffs of the Huachucas seemed especially effective at bouncing the signals. This made it difficult to tell exactly where the signal was coming from, and in some parts of the canyons a strong signal would seem to be coming from both sides of the canyon. It took a bit of experience and scrambling up and down cliffs to distinguish direct signal from bounce in these situations. If the coatis were up on the ridges or in the upper cliffs, we could often get a signal from the base of the mountains. If they were on the bottom of the canyons, sometimes we had to be almost on top of them to get a signal. In addition, the Army Intelligence Center at Fort

Huachuca was constantly testing communication systems, and it was quite common to receive a lot of interference that affected the ability to receive the low-energy broadcasts from the radio collars. This interference also meant that radio tracking from a plane or helicopter was not practical.

The technology of the time, VHF collars, required a lot of leg work to find the collared coatis. I was lucky enough to have one paid assistant for the first 16 months of the study and several part-time volunteers off and on during the rest of the study. We tried to find each troop at least once per week, and when possible, sneak up on them to observe behavior and count the number of animals in the group. But all that legwork, climbing over hill and dale trying to find coatis, also provided an intimate connection with the landscape that would not have been possible otherwise. Following the coatis on an almost daily basis for years allowed us to absorb and appreciate not only the rhythms of the coatis' lives, but the rhythms of the mountains. Observing the changing of the seasons, through winter snows, the dry and windy springs, the lushness of the monsoon, and the crispness of fall, added a necessary dimension to understanding how coatis make a living in these mountains. It takes several years to appreciate this rhythm and see the subtle and sometimes not-so-subtle differences from year to year - differences, it turned out, that have profound impacts on coati population dynamics and behavior. Overall, the effort was worth it. My study yielded many new insights into how coatis survive (or not) in these northern parts of their distribution and provided intriguing comparisons to studies done in other areas.

Other research on coatis

When I first began pursuing the idea of a coati study in the early 1990's, coatis were relatively unstudied. John (Jack) Kaufmann's and Jim Russell's research on Barro Colorado Island in Panama had been the primary studies.[6–9] Several studies attempted in Arizona had been short-lived, due to low population sizes and the difficulty in following animals that ranged over large areas. White-nosed coatis became a fashionable topic for studies in the early 1990's, with studies in southeastern Arizona, Mexico, Guatemala, Costa Rica, and Panama. Additional work was occurring in the lab, museum, and zoo, concerning learning and visual abilities, systematics, and communication.[10–16] The first studies of the ecology of brown-nosed coatis began in the mid-1990s, mostly in Brazil and Argentina.

Very few studies have been conducted on mountain coatis,[17,18] with the only published field studies conducted by Rodriguez-Balaños and colleagues on the food habits, activity and home range of western mountain coatis in Columbia, during the late 1990s.[19,20]

These studies in different parts of South America have greatly expanded our knowledge of what a coati is, even if we refer to them as different species. Although these studies used similar techniques to mine, most of them were done on habituated animals, providing the researchers with much more than the fleeting glimpses of coatis I considered myself fortunate to get. These studies have supplied some of the most interesting observations of the intimate details of coati lives, but in many cases, the habituated animals were also food supplemented. What affects this added resource has on coati behavior or ecology has not been addressed yet.

Notes and References

1. Anesthesia was done under supervision of the Fort Huachuca veterinarians.

2. Gehrt SD, Hungerford LL, Hatten S (2001) Drug effects on recaptures of raccoons. Wildlife Society Bulletin 29:833–837

3. Thanks to S. Ratneyeke for the plunger design (via S. Stone).

4. For more about causes of coati mortality, see chapter 4.

5. Manufactured by Advanced Telemetry Systems, Isanti, MN, USA.

6. Kaufmann JH (1962) Ecology and social behavior of the coati, *Nasua narica*, on Barro Colorado Island, Panama. University of California Publications in Zoology 60:95–222

7. Russell JK (1981) Exclusion of adult male coatis from social groups and protection from predation. Journal of Mammalogy 62:201–206

8. Russell JK (1982) Timing of reproduction by coatis (*Nasua narica*) in relation to fluctuations in food resources. *In*: Leigh EG Jr (ed) The ecology of a tropical forest. Seasonal rhythms and long-term changes. Smithsonian Institute Press, Washington, D.C., pp 413–431

9. Russell JK (1983) Altruism in coati bands: nepotism or reciprocity? In: Wasser SK (ed) Social behavior of female vertebrates. Academic Press, New York, NY, pp 263–290

10. Chausseil M (1991) Visual same-different learning, and transfer of the sameness concept by coatis. Ethology 82:28–36

11. Chausseil M (1992) Evidence for color vision in procyonides: Comparison between diurnal coatis (*Nasua*) and nocturnal kinkajous (*Potos flavus*). Animal Learning and Behavior 20:259–265

12. Decker DM (1991) Systematics of the coatis, genus *Nasua* (Mammalia: Procyonidae). Proceedings of the Biological Society of Washington 104:370–386

13. Decker DM, Wozencraft WC (1991) Phylogenetic analysis of recent procyonid genera. Journal of Mammalogy 72:42–55

14. Compton LA, Clarke JA, Seidensticker J, Ingrisano DR (2001) Acoustic characteristics of white-nosed coati vocalizations: a test of motivation-structural rules. Journal of Mammalogy 82:1054–1058

15. Maurello MA, Clarke JA, Ackley RS (2000) Signature characteristics in contact calls of the white-nosed coati. Journal of Mammalogy 81:415–421

16. Shannon D, Kitchener AC, Macdonald A (1995) The preputial glands of the coati, *Nasua nasua*. Journal of Zoology, London 236:319–357

17. Balaguera-Reina SA, Cepeda A, Gonzalez-Maya JF (2009) The state of knowledge of western mountain coati *Nasuella olivacea* in Colombia, and extent of occurrence in the northern Andes. Small Carnivore Conservation 41:35–40

18. Helgen KM, Kays R, Helgen LE, Tsuchiya-Jerep MTN, Pinto CM, Koepfli K-P, Eizirik E, Maldanaldo JE (2009) Taxonomic boundaries and geographic distributions revealed by an integrative systematic overview of the mountain coatis, *Nasuella* (Carnivora: Procyonidae). Small Carnivore Conservation 41:65–74

19. Rodriguez-Bolaños A, Cadena A, Sanchez P (2000) Trophic characteristics in social groups of the mountain coati, *Nasuella olivacea* (Carnivora: Procyonidae). Small Carnivore Conservation 23:1–6

20. Rodriguez-Bolaños A, Sanchez P, Cadena A (2003) Patterns of activity and home range of mountain coati, *Nasuella olivacea*. Small Carnivore Conservation 29:16–19

PART TWO
NATURAL HISTORY

4

Birth

I call April, May, and June in the Huachucas the "dead season," as life seems to wither and disappear, parched and dried up, hiding from the searing sun, waiting and waiting for the onset of the summer rains. The grass fades to pale yellow; the hillsides are spotted with brown then gray as the leaves of the Madrean oaks change color and fall to the ground. The creeks and springs dry up. Week after week the weather is clear, hot, dusty dry. Trying to sneak up on coatis is nearly impossible, as the leaf litter becomes crispy as potato chips. In late June, sometimes early July, with temperatures in the mountains nearing 38°C (100°F), the first thunderheads start to form. Dry lightning, with little rain, accompanies these first storms, often triggering scattered wildfires. The singing of the cicadas reverberates through the woodland so loudly that you feel like your head will explode. It is near the end of this hot dry season when the coatis give birth.

The Groups Start to Split Up

Following the mating season in late March to early April, the troops start to splinter and small subgroups form. In mid-June, pregnant females start wandering on their own as they begin searching for a nesting site for their new litter. Most female coatis seek isolation when they give birth, leaving their earlier offspring and other troop mates behind. A female coati's behavior changes markedly when she leaves her troop. By a week or two before she gives birth, she is massively pregnant, looking like she has swallowed an entire watermelon. She moves ponderously, often stopping to pant in the heat. And no wonder – female coatis give birth to an average of four young, each weighing about 114 to 117 gm (4 to 6 oz), for an average litter weight of 455 to 680 gm (1 to 1 1/2 lbs), a whopping 11-17 percent of her body weight![1-4] This is the equivalent of a 220-kg (100-lb) human female giving birth to an 5-8 kg (11-17 lb) baby - not a record breaker, but above the 95th percentile of birth weights for humans. In contrast, similar-sized raccoon mothers bear a litter weight less than half that of coati mothers.[4]

So, female coatis face a heavy neonatal burden for an animal that not only has to lug around the extra weight, but also needs to be nimble enough to flee from predators – indeed three radiocollared females were killed by mountain lions within a couple of weeks before they would have given birth.

Nesting Behavior

When female coatis leave their troops, they spend several days looking for and constructing a nest. Coatis do their best to hide the locations of their nests, but I managed to find 27: seventeen were in rock crevices, three were in caves or sinkholes, three were in tree cavities, and one was under a horse barn. Some females used the same nest in subsequent years, but most chose a new location each year. They usually nested alone, but in at least one case, two females shared a sinkhole (cave), although it is possible that their nests were in different portions of the underground cavern. In Guatemala and Panama, a few female coatis were observed sharing nests, and they were even observed nursing each other's kits.[5,6] In most areas of the tropics, coatis build elaborate covered platform or spherical nests high in the forest canopy,[2,5,7,8] but the coatis in the Huachucas used cavities in the trees, either in oaks or sycamores, rather than platforms. Brown-nosed coatis at Iguazu National Park in Argentina also used crevices among rocky cliffs.[9] Coatis in the Huachucas often built and used small tree platforms for nighttime resting during the rest of the year, but I never observed a female to use a platform to house her babies. However, coatis that live along riparian drainages in Arizona, with stands of large cottonwood trees and few rocky crevices, have been observed nesting in platforms high in the trees.[10]

Finding coati nests was a real challenge – for two reasons. First was the activity of the mothers after they gave birth: they left the nest early in the morning, and often did not return until late afternoon, so it was hard to catch them in their nests. Second, the mothers were crafty about keeping the locations of their nests hidden. I spent many long days carefully sneaking up on radiocollared females that I thought might be in their nests, only to find them on the ground or in the low branches of a tree watching me trying to be sneaky. They quietly sat, often allowing me to approach and take photos before they slow wandered off, usually in the opposite direction of the suspected nest location.

This "calm" behavior of females was characteristic when they had babies in a nest. Females that used to flee as soon as they detected a human now could be approached quite closely. I heard other people also describe this behavior, saying that the coatis seemed unconcerned by their presence. I do not think it was a lack of concern at all. I think the females use this behavior to keep an eye on danger and to lead potential predators away from their nest. I termed it "decoy" behavior. Two radiocollared females were killed by mountain lions and one by a black bear shortly after they gave birth, and I often wondered if they might have been using this decoy behavior to try to lead the predators away from their nests, perhaps misjudging their ability to escape at the last moment.

The Babies Are Born

Once a female gave birth, she did not leave the nest for a day or so, and in some cases that allowed us to get the nest location. From then on, she would usually leave the nest mid-morning and spend most of the day foraging. She would return in late afternoon, nurse the kits and rest awhile. Sometimes she would go out for a bit more foraging just before dark. Then she would spend most of the night in the nest with the kits.

Approaching a female coati in a nest full of kits had to be done carefully. My assistant, Jason, found this out the hard way. Late one afternoon in July 1996 we climbed up some rocky ledges in Garden Canyon where we suspected two different females had their nests. The afternoon thunderstorms had already come and gone, and the sky was clearing as sunset approached. The female I chose to follow, F35, (the designation "F" referring to female and the number being the eartag number) led me up over the top of the ridge and into another canyon, and darkness descended before she

F33 watching me trying to sneak up on her to determine the location of her nest, July 1997.

returned to her nest (possibly another example of decoy behavior). Jason followed the signal of F33 to a small crack in a layer of limestone. He shimmied his way up the rock and peered into the crack. Just then F33 charged, growling and snarling, her eyes lit up by the setting sun. Jason released his grip and dropped the 10 feet to the ground. When I caught up with him a few minutes later, he was white as a sheet, and I bet he still has nightmares about the snarling coati with the glowing red eyes.

During 1998, Michael Sutor, a film maker from Germany, contacted me about filming the coatis for a documentary he was making about brown-nosed and white-nosed coatis. I tried to dissuade him from using my animals, as I thought that they would be too difficult to approach for filming. But he came anyway and, to my surprise, was quite successful in finding and filming coatis. With luck and determination, and aided by the telemetry gear, he was able to get some good footage. And, in a moment of great fortune, he managed to watch one of the radiocollared females, F149, as she was setting up a nest in a cavity in an oak tree, not far from a dirt road in Garden Canyon. A couple of days after she gave birth, when F149 was out foraging, he used a ladder to access the nest and installed a small camera that, via a lengthy cable, fed into the video recorder in his vehicle. He was able to get footage of the F149 and her five kits, which appeared in the film, *Nasenbären*.[11] I never attempted to peek into any coati nests, as I was too afraid of disturbing the youngsters or causing the mother to move them to a new nest. I am indebted to him for providing me with intimate details of newborn coatis that I otherwise would not have been able to obtain.

Coatis are born blind and helpless, covered in a thin layer of soft gray fur. They are rather grotesque when they are born, with a large head and bulbous nose that looks way too large for their body. Their eyes open in a week to 10 days, about the time they are able to walk, and their teeth start appearing soon after.[1,12] And then their nose starts to lengthen. Early days are spent exploring their new world, and this soon evolves into play with their litter mates. They bite, wrestle, and pounce. They grow quickly, and by the time they are ready to leave the nest they have increased their weight from about 100 gm (4 oz) to more than 400 gm (1 lb).[13]

I was able to determine the actual number of kits born for only one litter (F149's). The remaining litter sizes were based upon how many youngsters the mothers brought out of the nest. Once the females joined up with other females and their offspring, which they often did soon after leaving the nest, it was almost impossible to figure out which kits belonged to which female. So, I monitored each female carefully and tried to see her with her youngsters before she rejoined the troop. Females brought up to six kits back to the troop, with an average of four.[14] There was a tendency for older females to return with slightly larger litters (2-3 year olds: mean = 2.9, 4-6 year olds = 3.7, 7 + year olds = 4.1).[15] Of the females of known age, two gave birth to their first litter when they were two years old, and one waited until she was three. Rarely, females returned to the troop without kits: two females from Troop 5 in 1997 nested and stayed near their nests for the usual six weeks but rejoined the troop alone. All other females that were at least three years of age gave birth and brought kits back to their troops each year.[16] These findings are similar to those further south, although some female coatis in Panama did not reproduce every year, especially when they were older.[2,3,5] In Costa Rica, some females lost all or part of their litters to predation by white-faced capuchins.[8,9,17-19] Eagles and other raptors have also been observed preying on baby coatis.[5,20] Capuchins don't live in Arizona, and overall, mortality of baby coatis in the nest was quite low.

During 1999, F185 gave birth under a horse barn in Ramsey Canyon. In early August, her radio signal was coming from inside a shed located on Nature Conservancy property in the lower part of the canyon. At the time, The Nature Conservancy also kept an office across the street, where naturalist Mark Pretti had a desk. I stopped by to talk with him, and he mentioned that someone had seen F185 lead her wobbly youngsters into the shed. My truck was parked near the shed, so I watched from the cab of my truck to see if F185 would bring her kits out, allowing me to get a count. She appeared alone and took off along the creek. Grabbing an empty ice chest from my truck and enlisting Mark's assistance, we carefully entered the shed, which was being used to store large garbage bags full of leaves that had been raked up around the office. I could not see the kits but could hear them against the far wall. I crawled over the bags of leaves, hoping there were no rattlesnakes or scorpions hiding about the bags, until I found the stash of kits. One by one I carefully picked up the youngsters and handed them to Mark who placed them in the ice chest. When all five had been captured, we quickly took our prize across the street to weigh and measure them. Although we were very gentle, the babies bit, clawed, squealed, and squirmed. We were able to get weights, but few other measurements as I wanted to make sure we got them back to their hiding spot before F185 returned.[13] When we had weighed them all, we quickly returned them to the shed and carefully placed them back where we found them. And just in time; F185 returned within five minutes and just a few moments later she led the youngsters out of the shed. When I was next able to see her (she spent a lot of time on private property in the canyon where I did not have access) a couple of weeks later, she still had her kits with her.

Female coatis in Arizona gave birth during a brief period each year, usually a span of less than 10 days for all the females in the population. Exact birth dates were difficult to figure out, but by observing obviously pregnant females in the areas of their nests, then observing them a few days later looking much skinnier, I could estimate within a day or so of when they gave birth. With one possible exception, all births occurred between June 14 and July 1[21] The average birth dates differed by up to several days from year to year.[22] These were only the birth dates for radiocollared females, but I was unaware of any other births from marked or unmarked coatis outside of this time. The one possible exception was F167. She was trapped and radiocollared on July 16, 1998. I had first captured her in January of that year and estimated her then at 1½ years of age, based on overall body size and tooth wear. When I captured her again in mid-July, she had already given birth, her first litter, and was obviously lactating. I was able to track her to her nest and monitor her closely. She did not bring her kits back to the troop until almost the end of August, about two weeks after the other females had already rejoined. Her kits appeared to be about six weeks old, the same as most kits when brought out of the nest. This would put the date of the birth of the kits around the second week in July. It's not unusual for young female mammals to conceive slightly later than older females, especially with their first litters,[23] but F167 was the only two-year-old for which I had even an approximate birth date, so I do not know how common that might be for coatis. Births within a troop were usually within just a few days of each other, but there might be up to 13 days difference in average birth dates between different troops. This pattern, of a brief period of births that differs slightly among troops and among years, has also been observed in Guatemala, Panama, and Brazil.[2,5,9]

Emerging from the Nest

Baby coatis spent about six weeks in the nest, with some females bringing their youngsters out in as little as 32 days, and others waiting as much as 47 days. The dates that females brought their kits out of the nests were not as clustered in time as were birth dates. Emergence dates of kits of marked females were usually spread over about two weeks, with several additional weeks for all the females to find each other and rejoin as a troop. The earliest I saw kits appear from the nest was July 21, and the latest August 28. The average date was August 3 and did not differ much from year to year.[24]

The location data from the radiocollared females showed that they stayed close to their nests during the nesting period – usually within 200 m (0.1 mi) - and foraged within about a 290-hectare (700-acre) area.[10,25] Right before they brought their youngsters out of the nest, coati mothers increased their movements substantially, suddenly venturing over 1 km (0.6 mi) from the nest, and they appeared to be searching for their troop

F33 with her kits shortly after she brought them out of the natal den. She had four kits with her, and two are visible here (one behind her tail). She was aware of my presence, but had not yet discovered where I was, and is exhibiting alarm piloerection.

mates. I commonly saw the females foraging with another female or a few subadults within a few days before they brought their kits out of the nest.

However, reaggregation seemed to progress in fits and starts. One day a female and her kits would be in the company of another female with kits from her troop and then next day they would be separate, foraging only with their own kits. Sometimes several troop members would forage together for a few days, only to split up again. At other times, the troop females would bring their kits all together within a couple of weeks of exiting the nests and stay together until the following spring. It was often late September before the whole troop, including subadults, were all together again.

When I was not following females around trying to find their nests, I was watching nests, to figure out the activity patterns of the youngsters and coati mothers. A few nests were either close enough to roads, so I could watch the nests from my truck, or in an open enough spot so I could watch from an opposite hillside. I spent many hours staring at the nest entrances with binoculars to see if any babies would poke their heads out, all the while monitoring the telemetry receiver to hear when their mother

Left, 7- week-old white-nosed coati, Huachuca Mountains, 1997. Right, slightly older brown-nosed coati, Iguazu National Park, Argentina. Photo by Ben Hirsch.

was returning. For these few monitored nests, once the youngsters appeared capable of getting out of the nest, their mothers moved them. And when the kits left the nest, they left for good. Unlike many other animals that will take forays from the natal nest and return before finally dispersing, once the kits were able to follow their mothers, they moved to new nest sites. And as I will discuss in an upcoming chapter, coatis typically change nest sites every few days. Once the kits were on the ground, the weaning process started in earnest. The youngsters started eating solid food immediately, digging through the leaf litter just like the big coatis. They often approached and foraged right next to their mother, and I even saw them sniffing her mouth, which may be a way for them to learn what is edible. They continued to nurse for a couple of months, not only from their mothers, but occasionally from other lactating females in the troop. The kits quickly grew strong, and within a week or so had no trouble keeping up with the adults, especially when fleeing from researchers trying to sneak up on them.

Notes and References

1. Gilbert B (1973) Chulo. A year among the coatimundis. Alfred A. Knopf, New York, NY
2. Kaufmann JH (1962) Ecology and social behavior of the coati, *Nasua narica*, on Barro Colorado Island, Panama. University of California Publications in Zoology 60:95–222

3. Valenzuela D (1999) Efectos de la estacionalidad ambiental en la densidad, la conducta de agrupamiento y el tamaño del área de actividad del coatí (*Nasua narica*) en selvas tropicales caducifolias. Ph.D. Disertación, Instituto de Ecología, UNAM, México, D.F.

4. Whiteside DP (2009) Nutrition and behavior of coatis and raccoons. Veterinary Clinics of North America: Exotic Animal Practice 12:187–195

5. Binczik GA (2006) Reproductive biology of a tropical procyonid, the white-nosed coati. Ph.D. Dissertation, University of Florida, Gainesville

6. Russell JK (1983) Altruism in coati bands: nepotism or reciprocity? *In:* Wasser SK (ed) Social behavior of female vertebrates. Academic Press, New York, NY, pp 263–290

7. Olifiers N, Bianchi RC, Mourao G, Gompper ME (2009) Construction of arboreal nests by brown-nosed coatis, *Nasua nasua* (Carnivora: Procyonidae) in the Brasilian Pantanal. Zoologia 26:571–574

8. Sáenz JM (1994) Ecología del pizote (*Nasua narica*) y su papel como dispersador de semillas en el bosque seco tropical, Costa Rica. Master's Thesis, Universidad Nacional, Heredia, Costa Rica

9. Hirsch BT (2007) Within-group spatial position in ring-tailed coatis (*Nasua nasua*): balancing predation, feeding success, and social competition. Ph.D. Dissertation, Stony Brook University, Stony Brook, NY

10. Ratnayeke S, Bixler A, Gittleman JL (1994) Home range movements of solitary, reproductive female coatis, *Nasua narica*, in south-eastern Arizona. Journal of Zoology, London 223:322–326

11. "*Nasenbären*" (in German) was released in 2000 by the German Public Broadcasting Service.

12. Smith HJ (1980) Behavior of the coati (*Nasua narica*) in captivity. Carnivore 3:88–136

13. Data from one litter in the Huachucas, weights of 450, 400, 450, 400 and 420 g the day after they appeared from the nest. Also, Valenzuela (1999); Kaufmann (1962).

14. N = 34, standard deviation = 1.67.

15. ANOVA $F_{2,32}$ = 1.91, P = 0.1641.

16. 94% of 31 adult females.

17. Fedigan LM (1990) Vertebrate predation in *Cebus capuchinus*: meat eating in a neotropical monkey. Folia Primatologica 54:196–205

18. Rose L, Perry S, Panger MA, Jack K, Manson JH, Gros-Louis J, Mackinnon KC, Vogel E (2003) Interspecific interactions between *Cebus capucinus* and other species: data from three Costa Rican sites. International Journal of Primatology 24:759–796

19. Newcomer MW (1985) White-faced capuchin (*Cebus capucinus*) predation on a nestling coati (*Nasua narica*). Journal of Mammalogy 66:185–186

20. Hass CC, Valenzuela D (2002) Anti-predator benefits of group living in white-nosed coatis (*Nasua narica*). Behavioral Ecology and Sociobiology 51:570–578

21. Known females, 1996: June 20 (1 female), June 24 (2), June 27 (2), July 1 (2). 1997: June 20 (1), June 22 (3), June 26 (1). 1998: June 14 (1), June 16 (1), June

Birth

19 (1), June 26 (1) and June 28 (1). 1999: June 24 (1), June 28 (1), June 29 (1), June 30 (2). 2000: June 23 (1), June 26 (2).

22. Oneway ANOVA of Julian birth dates against year, $F_{4,22}$ = 5.09, P = 0.0064.

23. Bronson FH (1989) Mammalian reproductive biology. The University of Chicago Press, Chicago, IL

24. 1996: mean emergence date, August 1; 1997, July 31; 1998, August 8; 1999: August 6; 2000: August 5. Oneway ANOVA of Julian emergence dates against year, $F_{4,23}$ = 2.01, P = 0.1271.

25. Hass CC (1997) Ecology of white-nosed coatis in the Huachuca Mountains, Arizona, a preliminary study. Final report submitted to the Arizona Game & Fish Department, Phoenix, AZ. 1–52

5

Timing of Reproduction

Following the hot, dry spring and early summer, the monsoons begin with a change in wind direction and the appearance of the first clouds near the mountains. Mornings begin calm and crystalline. By late morning, on most days, cumulonimbus clouds start to build over the mountain tops. Sometimes in the late morning, but more often in the afternoon, thunder echoes through the canyons, hollow and haunting at first, becoming louder and more metallic as the storm nears. The rain starts as scattered, large droplets, quickly followed by a downpour, as the thunder causes the whole mountain to shake. I make a mad dash to my pickup – telemetry gear is quite the lightening attractant – and wait out the storm while catching up on notes and maybe taking a nap. The storms usually pass quickly but can dump several inches of rain in less than an hour, turning rivulets into waterfalls and creeks into torrents.

Annual Timing of Births

Nowhere is the relationship between coatis and the monsoon as clear as in the timing of reproduction. Throughout the range of coatis, from Arizona to Argentina, baby coatis are born at the start of the rainy season, which begins during late spring to early summer. At first, this appears to make no sense. Why would the adult females suffer through late gestation, an important period for neonatal development, during the dry season when food is least available? The most likely explanation is that the onset of the warm rains is followed by a flush of insects – high quality food for young coatis being weaned from their mother's milk. This pattern of timing reproduction so that growing youngsters have access to the highest quality food as they are being weaned is common among mammals.[1-3] Even though the mating season may involve a tremendous expenditure of energy to find and defend mates, and gestation may double a female's nutritional requirements, the energy needed by youngsters during weaning is ultimately the most important in determining when to mate.[1,2,4]

Several studies in the tropics have found that leaf-litter invertebrates[5] (a major portion of the coati diet) are much more abundant during the wet season than the dry season.[6-8] Likewise, in the Huachucas we found that the invertebrates that appeared most often in the coati diet were also most abundant during the summer rainy season (see Chapter 13). Gerald Binczik, studying coatis in Guatemala, hypothesized that seasonal reproduction in coatis was tied to the rains, and the ensuing increase of not only the availability of invertebrate prey, but also the protein content, which peaked in the middle of the wet season. This would mean that baby coatis were weaned off their mother's milk just as a high-protein alternative source of food became available. Fruits are also an important food for coatis and supply an excellent source of energy and calories, but they are usually lower in protein than invertebrates. Fruits, especially in the tropics, are sensitive to periodic droughts,[9] and reductions in fruit availability due to droughts have been implicated in several temporary population declines of coatis.[8,10,11] Invertebrates appear to be not only higher quality food for growing coatis, but also more predictable in time.[6]

The rainy season in Panama begins in late April and lasts until December. The rains move northward, arriving in central Mexico in early June and in Arizona in early July. In Panama, coatis give birth in April and May and bring their young out of the nest in May to June.[11] The timing in Costa Rica and Guatemala is similar.[6,12] In Jalisco, on the west coast of Mexico, births occur in June and July, just as the rains begin.[13] On the east coast of Mexico, baby coatis were on the ground by mid-June, indicating births around late April.[14] And in southern Arizona, births occur during the last two weeks of June and first week in July, and kits appear on the ground in late July and early August.[15,16]

Brown-nosed coatis also give birth with the start of the rains. In the southern hemisphere, the rains begin heading south of the equator during August, arriving in southeastern Brazil by October and lasting six to nine months. Few populations have been studied as intensively as have white-nosed coatis, but in studies done in Brazil and Argentina, between latitudes 10º and 26º S, coatis give birth just after the start of the rainy season in October and November.[17-21]

There are a couple of striking features of coati reproduction. The first is the timing, with populations throughout their range adjusting the timing of births to the start of the warm rainy season. For the six populations of white-nosed coatis for which data are available, in all but one births occurred just before the rains began, so that conditions were lush and food was abundant when the kits climbed out of the nest. Note that latitude alone does not explain the differences in the timing of birth seasons. For example, although Tamaulipas is farther north than Jalisco, the rains begin in May, while in Jalisco they do not usually begin until June, and the timing of coati births reflected that. For five studies of brown-nosed coatis that published estimated birth periods, all births occurred shortly after the rains began, but few data are available on differences among groups or among years.

As mentioned earlier, for my study site the timing of births differed a little among years and among troops. This has also been found for the other study sites, although for most sites the data are not very extensive, that is, only a few births were recorded, and the populations were only studied for short periods. The rainy season begins at different times throughout the coatis' range, and in addition, mating takes place about 70-75 days before, during the dry season. This means that coatis must be able to predict when the rains will come in each area. They cannot wait until the rains begin – the babies might be weaning just as the rains end and nutrients are harder to come by.

Most long-lived animals in temperate regions breed seasonally, having an internal clock that signals the time of year. Laboratory studies of a variety of animals have shown that this internal clock functions by recording the light through the retina of the eye, which sends a signal to the pineal gland to stop the release of melatonin.[22,23] Melatonin is released in darkness, and the duration of melatonin release tells the animal the length of the night, or inversely, the length of the day, or photoperiod. Longer days result in shorter periods of melatonin release, shorter days mean longer periods of melatonin release.[1,24,25] The repeated exposure to light and dark cycles establishes an approximate 24-hour circadian rhythm.[1,25] For seasonal breeders, it is usually not the absolute amount of day versus night, but rather the change in the amount of dark from day to day that tells an animal what time of year it is. Increasingly longer nights indicate the onset of fall and winter, whereas increasingly shorter nights indicate the onset of spring and summer.[26]

Although circadian clocks have been widely studied, much less is known about how organisms measure longer time spans. A wide variety of mechanisms have evolved to make sure animals time reproduction to ensure survival of their offspring.[1,27] In addition to being able to detect the time of year, some animals have an interval timer that measures the amount of time that lapsed since day length changed by a certain amount; others use a circannual (or, about a year) clock. Both interval timers and circannual clocks need periodic exposure to daylight and/or temperature to fine tune them and maintain schedules over long periods of time. Circannual clocks are particularly useful for animals that hibernate and go long periods without exposure to environmental cues.

At the equator, photoperiod changes little, if at all, throughout the year, making photoperiod a less useful clue for tropical animals. So how do coatis predict when the rains will come? At the southernmost studied population of white-nosed coatis, in Panama (9° North), daylight changes only a few seconds per day during December, when coatis are getting primed for mating in January and February. Is that small change enough to tell coatis what time of year it is? Does the 30-second change per day during the preceding October tell them the season is changing? We do not know the answer yet, but photoperiod is the most reliable environmental clue available to coatis, as the onset and amount of the rains are variable from year to year. But the difficulty in detecting that minor change might be revealed in the year-to-year variability in birth dates in that study site. Kaufmann, during his study from 1958 to 1960, estimated that births occurred during early to mid-May during 1959 and early April during 1960.[11] Russell, studying the same population from 1975 to 1978, recorded births only in early May.[8] Could a birth period extending at least six weeks among years be due to the inability to track changes in daylight very closely?

Birth Synchronicity

The second interesting feature of coati reproduction is the synchronicity of births.[27] The only other studies, besides mine 31° N), to focus closely on reproduction in coatis were Binczik's study in Guatemala (17° N), and

Ben Hirsch's study of brown-nosed coatis in Argentina (26º S). Near the towering Mayan ruins of Tikal, Binczik was able to habituate several large groups of coatis and collect detailed reproductive data from 1994 to 1996. During 1995, the birth period extended from 16 to 29 April, and during 1996, from 22 April to 9 May, for a total span of 3½ weeks. He also recorded three females in 1995 and one female in 1996 that lost their litters, came into breeding condition again, successfully mated and raised second litters. This second birth period occurred during 30 July to 18 August 1995, and on 7 August 1996 – after the other females were already bringing their youngsters back into the troops.[6] In Argentina, Hirsch habituated several groups in Iguazu National Park during 2002-2004. All births occurred during October, and as with white-nosed coatis, births within a troop were within a few days of each other, but there were up to two weeks difference in birth dates among troops.[19]

In Arizona, the lush period during the rains lasts only from about mid-July to mid-September. Farther south, however, the rainy season lasts longer – up to 7 months in Panama and Costa Rica, and 10 months in Argentina. Yet in each population, all births occur in a short time span, usually within a week or two of each other. However, in both Costa Rica and Guatemala, a second period of births was recorded for a few females that had lost their first litter, with females giving birth about four months after the first birth period and bringing their kits back to the groups in late August or September. Binczik recorded that several late-born youngsters survived more than six months in his study area in Guatemala.[6] This second period of births indicates that both females and males maintain the ability to mate; that is, males are producing viable sperm, and females are able to ovulate, from early February until late June. Among a captive group of coatis in Arizona, which presumably came from Arizona or northern Mexico, females that lost their young shortly after birth exhibited signs of a possible repeat estrus.[28] However, given the short rainy season in Arizona, it is unlikely that youngsters born from a second estrus period would survive.

During my coati study, I collected blood samples to examine reproductive steroids for seasonal patterns. I sent the samples to Dr. Al Fivizzani at my alma mater, University of North Dakota, who was studying reproductive steroids in a variety of species. A preliminary run on 27 female samples for progesterone and estradiol were inconclusive, indicating a problem with sample collection or calibration during testing. A preliminary run on five male samples for testosterone looked promising, but before more samples could be run, the laboratory freezers were shut down when the Red River flooded during April 1997, inundating the town of Grand Forks, and causing wide-spread power outages and damage throughout much of the city.

I thought all was lost, however, a later run of male samples was not significantly different from pre-flood results, indicating that little degradation in the samples occurred when the freezers were shut down. A final run of 30 male samples showed that plasma testosterone levels were high (> 5 ng/ml) from November through April, with a peak during March – right at the start of the mating season.[29] Testosterone levels were lowest (< 2 ng/ml) during May through July, and the levels of testosterone in males < 3 years of age were lower than older males. Among seasonally breeding mammals, plasma testosterone levels often fluctuate annually, reaching highest levels when the majority of females are coming into estrus.[1,30,31] Binczik attempted to look at seasonality of reproductive steroids in his coatis in Guatemala, but a catastrophic freezer failure destroyed his samples.[6] A study of captive coatis in Brazil also found that testosterone levels peaked right at the start of the mating season, however, they found no seasonal differences in the presence of functional sperm.[32]

For part of his Ph.D. research, Binczik tested several hypotheses to explain the very tight birth synchrony among coatis: 1) coati birth synchrony minimizes juvenile mortality by promoting communal care of youngsters; 2) birth synchrony minimizes juvenile mortality by predator swamping; and 3) birth synchrony results from the coatis' reliance upon social mechanisms to enhance reproductive seasonality. These hypotheses are not mutually exclusive. In addition, hypotheses 1 and 2 address ultimate (evolutionary) causes, while hypothesis 3 addresses proximate (mechanistic) causes. He tested these hypotheses by recording birth and reaggregation dates for 28 females from three different troops during a period of two years.

Let me provide a little more detail on the theory behind each of these hypotheses before I reveal Binczik's conclusions. Under hypothesis 1, keeping the youngsters together, following reaggregation, allows all the females to provide care and watchful eyes for the youngsters. This form of communal care has been well documented in coatis,[8,11] but for this to be a driving factor in the evolution of birth synchrony in coatis, the groups should reform as quickly as possible, with time for reaggregation no longer than the duration of the birth period. That was not the case in Guatemala (nor in Arizona), so Binczik rejected this hypothesis.

The second hypotheses, predator swamping, is a common explanation for birth synchrony in ungulates, such as bighorn sheep and wildebeest. In essence, it is an odds game – if everyone produces their young at the same time, predators cannot kill them all, so some will survive. If this was an important factor for birth synchrony in coatis, Binczik predicted that the timing of the birth periods should be the same for all troops in an area. This was not the case in Guatemala, where females in different troops differed in average birth dates by as much as four days. Such a difference was not found in Arizona, but my sample sizes were smaller. However, in Hirsch's study, differences among troops of 1-2 weeks were observed.[19] Binczik rejected this hypothesis also.

The third hypothesis, that birth synchrony results from the social cues that coatis use to enhance seasonality, deserves some explanation. As I mentioned earlier, animals in the tropics may have trouble using photoperiod as a cue for seasonality due to the small changes in daylight from day to day. Most animals in the tropics that breed seasonally and synchronously are also very social; perhaps a few individuals may be sensitive enough to detect the change in photoperiod, and the others take their cues from them.[1,6] There is ample evidence of social mammals synchronizing reproductive cycles, typically using pheromones (think of the documented cases of female college roommates synchronizing their menstrual cycles). Binczik believed his data most supported this hypothesis, as it would explain the tight synchrony within a troop, but slight difference among troops. Binczik concluded that *"birth synchrony in and of itself is not adaptive in this species."*[6] He furthermore suggested that mating synchrony, not birth synchrony, was what was being selected for.

Hypothesis 3 explains how but not why coatis have such synchronous births. In addition, I have a couple of concerns about Binczik's interpretation of his results. First, although his sample sizes were adequate, he only had two years of data on one population. He was thus unable to examine differences among years or evaluate how other factors, such as maternal age, influenced birth dates. More than one factor could be producing synchronous births in coatis, and the importance of each factor may differ among populations or years. Also, implicit in Binczik's hypothesis testing is the assumption that the conditions that he observed represent the ones under which coatis evolved. We know the coati progenitor evolved in

North America, presumably when the conditions were more tropical, and genetic evidence has revealed that "modern" coatis probably arose in northern South America.[33] Indeed, during the Miocene tropical environments extended to high latitudes, where tropical animals might use photoperiod as a cue. Synchronous breeding was also found in Hirsch's study of brown-nosed coatis in South America (26° S).[19]

Based on the few studies of coati reproduction, synchronicity declines toward the equator. The coatis in Arizona appear to be most synchronous, while those on Barro Colorado in Panama appeared to be the least. This may also reflect the length of the rainy season – only 2-3 months in Arizona; as long as 9 months in Panama. Obviously, many more observations of coati births in different latitudes are needed to relate synchrony to length of rainy season. Birth data for coatis near or just north of the equator, where there appears to be no seasonality of precipitation, would be particularly interesting. Here, coatis would be exposed to year-round rain and only small changes in daylight. Future research should be directed toward detecting physiological mechanisms and evolutionary history, and should answer the hows and whys, including the role of mating strategies.

I propose another hypothesis for development of birth synchrony: it is beneficial to have the youngsters in the troops of similar age. Numerous studies of mammals have shown that baby mammals are more likely to play with similar-aged playmates.[34] During play, animals develop physical skills, but also develop social skills and lasting social bonds. They learn the rules for social engagement. They learn how to make and interpret subtle changes in body language that can indicate another animal's intent.[35-37] It may be within play that the various forms of cooperation seen in coati society have their root – this may be really stretching, as there have been few formal studies done on play in coatis. Observations of captive coatis indicate that they are very sensitive to rearing environments, with captive animals differing from wild coatis in social preferences and activity patterns.[11,38-40] But while it may be advantageous to have similar-aged playmates, would it make much difference if there were several weeks difference in age, versus several days?

Perhaps Binczik is correct, and synchronous births in coatis occur because some animals are cuing from others, rather than from slight environmental cues. Perhaps that is the best explanation at low latitudes, while there may be other explanations at higher latitudes. Coatis in the tropics may also use a circannual clock, sensing the time since the last rainy season, or the time since the summer solstice. In all likelihood, there are multiple explanations. But I would like to see more data on the precise timing and duration of birth seasons in more populations of coatis, across their range, before ruling out other hypotheses. In the meantime, we are stuck with the mystery of not only why they are synchronous in their breeding periods, but how.

Notes and References

1. Bronson FH (1989) Mammalian reproductive biology. The University of Chicago Press, Chicago, IL

2. Hass CC (1993) Reproductive ecology of bighorn sheep in alpine and desert environments. Ph.D. Dissertation, University of North Dakota, Grand Forks

3. Hass CC (1997) Seasonality of births in bighorn sheep. Journal of Mammalogy 78:1251–1260

4. Oftedal OT, Gittleman JL (1989) Patterns of energy output during reproduction in carnivores. *In*: Gittleman JL (ed) Carnivore behavior, ecology, and evolution. Cornell University Press, Ithaca, NY, pp 355–378

5. Michael Wilson, an entomologist friend, reminds me that the term "leaf-litter invertebrate" is vague and misleading, as it also includes microscopic organisms. In the context of coati food, "leaf-litter invertebrate" refers to insects, crustaceans, and mollusks at least 2 mm long.

6. Binczik GA (2006) Reproductive biology of a tropical procyonid, the white-nosed coati. Ph.D. Dissertation, University of Florida, Gainesville

7. Levings SC, Windsor DM (1982) Seasonal and annual variation in litter arthropod populations. *In*: Leigh EG Jr (ed) The ecology of a tropical forest. Seasonal rhythms and long-term changes. Smithsonian Institution Press, Washington, D.C., pp 355–387

8. Russell JK (1982) Timing of reproduction by coatis (*Nasua narica*) in relation to fluctuations in food resources. *In*: Leigh EG Jr (ed) The ecology of a tropical forest. Seasonal rhythms and long-term changes. Smithsonian Institute Press, Washington, D.C., pp 413–431

9. Wright SJ, Carrasco C, Calderon O, Paton S (1999) The El Niño southern oscillation, variable fruit production, and famine in a tropical forest. Ecology 80:1632–1647

10. Gompper ME (1997) Population ecology of the white-nosed coati (*Nasua narica*) on Barro Colorado Island, Panama. Journal of Zoology, London 241:441–455

11. Kaufmann JH (1962) Ecology and social behavior of the coati, *Nasua narica*, on Barro Colorado Island, Panama. University of California Publications in Zoology 60:95–222

12. Sáenz JM (1994) Ecología del pizote (*Nasua narica*) y su papel como dispersador de semillas en el bosque seco tropical, Costa Rica. Master's Thesis, Universidad Nacional, Heredia, Costa Rica

13. Valenzuela D (1999) Efectos de la estacionalidad ambiental en la densidad, la conducta de agrupamiento y el tamaño del área de actividad del coatí (*Nasua*

narica) en selvas tropicales caducifolias. Ph.D. Disertación, Instituto de Ecología, UNAM, México, D.F.

14. Caso A (1994) Home range and habitat use of three neotropical carnivores in northeast Mexico (*Felis pardalis, Felis yagouaroundi, Nasua narica*). Master's Thesis, Texas A&M, Kingsville

15. Personal observation.

16. Ratnayeke S, Bixler A, Gittleman JL (1994) Home range movements of solitary, reproductive female coatis, *Nasua narica*, in south-eastern Arizona. Journal of Zoology, London 223:322–326

17. Beisegel B de M (2001) Notes on the coati, *Nasua nasua* (Carnivora: Procyonidae) in an Atlantic forest area. Brazilian Journal of Biology 61:698–692

18. Costa E, Mauro R, Silva J (2009) Group composition and activity patterns of brown-nosed coatis in savanna fragments, Mato Grosso do sul, Brazil. Brazilian Journal of Biology 69:985–991

19. Hirsch BT (2007) Within-group spatial position in ring-tailed coatis (*Nasua nasua*): balancing predation, feeding success, and social competition. Ph.D. Dissertation, Stony Brook University, Stony Brook, NY

20. Olifiers N, Bianchi RC, Mourao G, Gompper ME (2009) Construction of arboreal nests by brown-nosed coatis, *Nasua nasua* (Carnivora: Procyonidae) in the Brasilian Pantanal. Zoologia 26:571–574

21. Resende BD, Mannu M, Izar P, Ottoni EB (2004) Interaction between capuchins and coatis: nonagonistic behaviors and lack of predation. International Journal of Primatology 25:1213–1224

22. Or more precisely, the signal from the retina stimulates the supra-chiasmatic nucleus of the hypothalamus, which sends a signal to the pineal gland that inhibits the release of melatonin.

23. Favaron PO, Mancanares CAF, De Carvalho AF, Ambrosio CE, Leiser R, Miglino MA (2008) Gross and microscopic anatomy of the pineal gland in *Nasua nasua* - coati (Linnaeus, 1766). Anatomy and Histological Embryology 37:464–468

24. Bronson FH (2009) Climate change and seasonal reproduction in mammals. Philosophical Transactions of the Royal Society of London B: Biological Sciences 364:3331–3340

25. Paul MJ, Zucker I, Schwartz WJ (2008) Tracking the seasons: the internal calendars of vertebrates. Philosophical Transactions of the Royal Society B: Biological Sciences 363:341–361

26. Some bacteria and plants also have circadian rhythms. Obviously, this is not done through pineal glands, but some other forms of photoreceptors.

27. Reiter RJ, Tan D-X, Manchester LC, Paredes SD, Mayo JC, Sainz RM (2009) Melatonin and reproduction revisited. Biology of Reproduction 81:445–456

28. Smith HJ (1980) Behavior of the coati (*Nasua narica*) in captivity. Carnivore 3:88–136

29. Hass CC (1997) Ecology of white-nosed coatis in the Huachuca Mountains, Arizona, a preliminary study. Final report submitted to the Arizona Game & Fish Department, Phoenix, AZ. 1–52

30. Davison LA (1993) Estimation of the peak breeding season for a raccoon (*Procyon lotor*) population based on fluctuations in serum testosterone, progesterone, and estradiol. Master's Thesis, Eastern Kentucky University, Richmond

31. Kaplan JB, Mead RA (1993) Influence of season on seminal characteristics, testis size and serum testosterone in the western spotted skunk (*Spilogale gracilis*). Journal of Reproduction & Fertility 98:321–326

32. Rodrigues de Paz RC, Santos Avila HB, Morgado TO, Nichi M (2012) Seasonal variation in serum testosterone, testicular volume and semen characteristics in coatis (*Nasua nasua*). Theriogenology 77:1275–1279

33. Nigenda-Morales SF, Gompper ME, Valenzuela-Galván D, et al (2019) Phylogeographic and diversification patterns of the white-nosed coati (*Nasua narica*): Evidence for south-to-north colonization of North America. Molecular Phylogenetics and Evolution 131:149–163

34. Fagen RM (1981) Animal play behavior. Oxford University Press, New York, NY

35. Bekoff M, Allen C (1998) Intentional communication and social play: how and why animals negotiate and agree to play. *In*: Bekoff, M Byers, J (eds) Animal play: evolutionary, comparative and ecological perspectives. Cambridge University Press, New York, NY, pp 97–114

36. Hass CC (1986) Play behavior and dominance relationships of bighorn sheep on the National Bison Range. The University of Montana, Missoula

37. Hass CC, Jenni DA (1993) Social play among juvenile bighorn sheep: structure, development, and relationship to adult behavior. Ethology 93:105–116

38. Gilbert B (1973) Chulo. A year among the coatimundis. Alfred A. Knopf, New York, NY

39. Mike Seidman, personal communication.

40. Stan and Linda Rolinski, personal communication.

6

The Social Carnivoran

When the troops reform in August, the Huachuca Mountains look entirely different than they did two months earlier when the pregnant females went off by themselves to give birth. Everything is green and flourishing. The evergreen oaks are sporting new foliage; grasses, forbs, and vines are everywhere. Mushrooms and ferns appear in the nooks and crannies of the limestone. It is hot and humid and, well, tropical. The ground is saturated from almost daily afternoon thunderstorms, and water is running in the streams. On the stream banks, chokecherries, canyon grapes, buckthorn berries, raspberries, sumac berries, and barberries are ripening. The leaf litter is crawling with beetles and their grubs, caterpillars, crickets, scorpions, and large desert centipedes. Baby frogs and toads occupy puddles and pools along the streams, and rattlesnakes – five species in the woodland and at least two more in the lower elevations – are on the prowl. Fawns of mule deer and white-tailed deer appear at the start of August. Rounding a corner while hiking in the mountains, the sight of a mossy rivulet surrounded by wildflowers makes me think I am in Montana and not southern Arizona.

Troop Structure

Into this soggy lushness appear the newest members of the coati population. For a brief period, I had radio collars in nine different troops centered on Fort Huachuca. These nine troops covered an area of some 13,431 ha (52 sq mi), over most of the northern half of the Huachucas.

The first coatis radio collared, in 1996, were from Troop 3. We captured seven coatis in two side canyons of Garden Canyon, about ½ km (¼ mi) apart, and put radio collars on two of the adult females. Two days later, at one of the same trap sites, four more females were captured, as it turned out, from the same troop as the first group.

Shortly after the second captures, the troop split into two (Troop 2 and Troop 3; Troop 1 was an often-observed troop in Ramsey Canyon that we monitored but did not radio collar until late 1996). Troop 2 included F27 and F34 and several unmarked individuals; a couple of weeks after being trapped in Garden Canyon, they headed to upper Ramsey Canyon. It was mating season, and, as I learned in later years, troops often split up during that period. F27 stayed in far upper Ramsey Canyon, a good distance south of the Fort Huachuca boundary, and gave birth in that area – 4.7 km (2.9 mi) from where she was captured. F34 was not radio collared so I only know her whereabouts from opportunistic sightings, but she was usually with F27. High fire danger closed the Coronado National Forest during June and July that year, but it reopened just in time to locate F27's nest before she brought her babies out in early August. In September, she returned to Garden Canyon and rejoined Troop 3, where she stayed until she disappeared three years later.

Coati troops are usually composed of 3-10 adult females and their offspring of the last couple of years, but overall group size varies over time and in different locations. In the Huachucas during my study, troop sizes averaged from 5 to 18 animals, and usually included 3-5 adult females, 2 to 8 juveniles (<1-year-old) and 1-3 subadults (1-2 years old). But for brief periods in the fall, after the females brought their young back to the troop and the subadults found their way home, troops could get quite large. It was not unusual then to see troops of 30 or more animals, mostly young kits. In some sites in the tropics or tropical dry forest, coatis split into subgroups for short periods (up to several days); these subgroups may include 5-10 animals, with 3-5 females.[1-4] Troops in the Huachucas were less likely to split into subgroups, but my observations may be biased by having too few radios in some troops. In other words, I could see the groups or subgroups that had radiocollared females in them, but I was likely to miss the groups that did not, unless I just stumbled across them. In one study in Guatemala, some troops were much larger, with one troop having 162 individuals, and did not appear to split into subgroups at all.[5] In Argentina, Hirsch reported groups of 8-65 brown-nosed coatis, and mountain coati groups in Columbia have been observed ranging from 6-8 to as many as 50 individuals.[6,7]

Young female coatis tend to stay in their natal troops, but males typically leave when they are approaching their second or third year. It is difficult to radiocollar young animals, as they are still growing and may outgrow their collars and strangulate. But in the few times it's been attempted, or where dispersal has occurred among recognizable animals, males typically establish home ranges near or overlapping their natal range.[8-10] This is what happened with M187. I captured and radiocollared him in November 1998 along with other members of Troop 5. I estimated his age, based on his large size and amount of tooth wear, at 2½ years. He stayed with the troop until February 18th, 1999. On that date, he was near the troop, but not part of it. From then on, he remained separate from the troop, spending much of his time just to the south, although his new home range overlapped Troop 1 and Troop 5. On April 30, he was located with Troop 5 for a few hours, and then he was not located with them again, but remained in a home range that overlapped Troop 5.

The troops in the Huachucas were remarkably stable over time, at least regarding group membership of adult females. I only know of two instances, involving three marked animals, of changes among troops. One of those involved F27 and F34, mentioned above, as they split from and later rejoined Troop 3. The other involved F83, originally captured as part of Troop 4 during June 1996. She was not radio collared but was eartagged and was occasionally seen in the company of radiocollared females F47, who died in early July 1996, and F79, who died of distemper in late December 1996. With F79's death, I lost track of Troop 4; given the virulence of canine distemper, I knew it was possible that the disease took out the entire troop. However, nine months later, I captured F83 again this time with the members of Troop 7. I observed her again in March of 1998, still with Troop 7. Groups splitting up and forming new troops, and animals migrating into troops have been reported in other study sites,[6,8,10,11] but no studies have lasted long enough to really assess emigration and immigration among troops for the duration of a coatis' lifetime (10 years or more, if they survive the gauntlet of predators and disease). Likewise, genetic work done in the tropics has shown that most but not all adult females within a troop are closely related.[8,12]

This brings up the question of how to define a troop? Troop home ranges overlap extensively (Chapter 14); there is some fission and fusion among groups, and some troops split into subgroups for short periods of time. So, the definition must be operational: a troop is a group of individuals repeatedly found together, in a distinct (but not necessarily exclusive) area. Troops are closely related genetically to neighboring troops, probably forming by division of adjacent troops, hence are, in essence, extended family groups.[10,12] This fission-fusion of female groups, in essence, "meta-families," has also been described for African elephants,[13] but appears to be rare among carnivorans. Using computer modeling, Avilés and colleagues demonstrated that occasionally allowing non-kin to join groups helped maintain critical group sizes without negatively impacting altruistic behaviors.[14]

The most stable and long-lasting troop was Troop 3 – the one I started with in 1996. I marked 16 animals in that group, of which at least five were still alive in early 2000. Some of the 16 were born during the study. Because of the large land area covered by the coatis, which was not known until we put radios on them, we could not maintain the same amount of monitoring effort on all the troops. Nevertheless, I trapped Ramsey, Garden, and Huachuca Canyons throughout the study, to keep some radios in each area. But most of my effort was focused on Garden Canyon, and my two most-studied troops (Troop 3 and Troop 5) included different parts of that canyon in their home ranges (see Chapter 14).

Determining the actual social structure of coati troops in the Huachucas turned out to be quite a computational chore, involving sifting through more than 3,000 locations.[15] In order to examine how much time any two individuals spent together, I had to determine how often I located each individual and how many times I located the pair together. Throughout the study, animals were continually being added to the monitored population (by trapping and marking) and leaving the population (through collar failure and mortality). In many cases, I could determine the location of the radiocollared animals from their signal alone without having to observe them. However, many of the animals were not radiocollared, but marked with only ear tags or dog collars or identifiable from unique markings, meaning that locations of those animals were strictly opportunistic, and usually in conjunction with radiocollared animals. If I could use a radiocollared coati's signal to stalk and observe the group, I might be lucky enough to see some of the other marked coatis as well. But

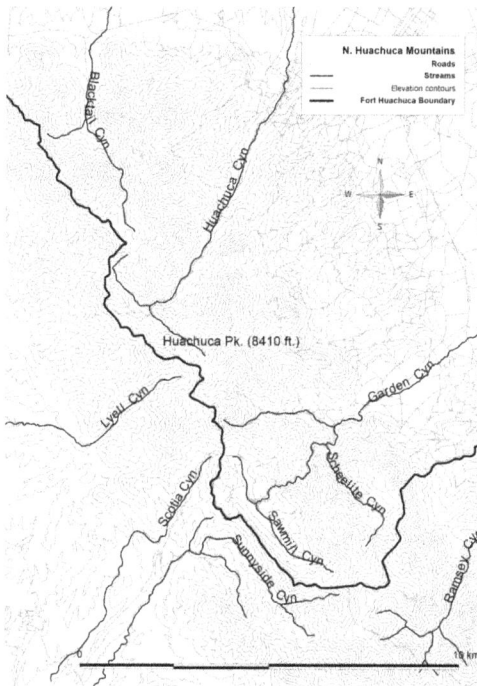

The study area in the northern Huachuca Mountains, indicating the locations of major canyons.

The study area in the northern Huachuca Mountains, indicating the locations of major canyons.

if the coatis were in thick vegetation, or they detected me and decided to flee, I might not see all the animals. That happened often. Because of the difficulty in knowing who was in a group, and not being able to be certain who was not in a group, I focused only on the animals with active radiocollars, hence could only look at troops that had more than one active radiocollar at a time (Troops 1, 3, and 5).

Troop 1 had two radiocollared females whose active radio times, when both females had functional radios, overlapped for 179 days. During this time, they were together on 93% of locations (not counting the nesting season, when females are solitary).[16] The 20 pairs[17] of radiocollared females in Troop 3 spent an average of 75% of their time together. But associations differed among pairs of females, with some found together as little as 40% of the time, and others as much as 96%. Again, this excluded the nesting season and only included the time periods when they could potentially have occurred together. The seven pairs of radiocollared females in

Troop 5 were together on average 61% of the time, with one pair located together only 29% of the time, and another 93% of the time.

Romero and Aureli found a distinct subgrouping among a group of captive brown-nosed coatis that shared the same enclosure. Although the females from the two groups occasionally interacted, most of the interactions were agonistic (e.g., lunges and chases), whereas most of the interactions within the subgroups were affiliative (e.g., mutual grooming, and nose touches).[18] Matt Gompper found that genetically-related females in groups of white-nosed coatis shared many more affiliative behaviors, whereas they were more antagonistic toward unrelated females in the group.[12] However, Hirsch found that aggression in the brown-nosed coatis he observed was not affected by relatedness.[19] It appears that social dynamics within troops are quite complex. At the time I did my study, the methods to determine relatedness among coatis were just being worked out, and I lacked the budget to have any genetic analysis done. But it would be interesting to know if the adult females were preferentially associating with close relatives. In another study in Arizona that overlapped temporally with mine, Maureen McColgin and colleagues found looser spatial relationships among three troops of coatis;[10] but the coatis were not observed year-round, and they might have defined a troop differently than I did.

Although home ranges of troops overlapped extensively, I rarely saw different troops together, and only for brief periods. Troop 3 and Troop 6 were located 312 times when both troops had active radio collars; only four times were they located together. Troop 3 and Troop 5 were located 694 times; only once were they found together. Likewise, Troop 6 and Troop 5 were only located once together. All these troops overlapped spatially and temporally. Troop 1 was near Troop 3 on several occasions, but they were never seen together. For example, I once observed them in Scheelite Canyon on a late winter afternoon. Troop 3 was spread out in the cliffs on the east side of the canyon, but as I was looking for marked animals in the troop, I noticed another group on the west side of the canyon with a radiocollared animal that did not match the frequencies of Troop 3. On a whim, I checked the frequency of the radiocollared animal from Troop 1, and sure enough, F258 and 10 other animals from Troop 1 were also in the canyon. But several other marked or recognizable animals from Troop 1 were not there, so it was a subgroup of Troop 1. Troop 1 crossed to the east side, headed up into the rocks and found a small crevice

Part of Troop 1 in Scheelite Canyon.

to den in for the night. Like some weird carnival trick, I watched in amazement as 11 animals disappeared into a crack that looked too small to hold even one coati. Troop 3 climbed up and around the crack where Troop 1 had holed up; one subadult approached the crack, only to be charged and growled at. The subadult quickly retreated and rejoined its group. Troop 3 moved higher up on the cliffs and disappeared. The next morning both groups were still in the canyon, but they remained separate until at least early afternoon when I left to find other coatis. Although both troops were in close proximity, I did not consider them "together."

I was lucky enough to be watching Troop 3 when Troop 2 rejoined them in September 1996, after splitting up the previous March. It appeared to be quite a joyful reunion, with several of the adult females grabbing each other's heads and grooming each other with their teeth – the coati version of a hug. Pretty soon both troops were all in one mass, a writhing ball of coatis as an orgy of mutual grooming ensued. With the addition of Troop 2, Troop 3 swelled from 31 to 45 animals, becoming the largest group I ever saw in the Huachucas. So, Troop 2 was temporary at best, and was probably a subgroup of Troop 3.

Spatial Structure of Groups

Most of a coatis' day is spent foraging for food, and much of that time foraging as a group through the leaf litter. When foraging, the troop spreads out in a rough ellipse.[6,11] Observing the entire troop foraging for any amount of time was nearly impossible in the Huachucas, due to dense vegetation and shy animals. However, a couple of studies in the tropics have provided more detailed information on the structure of foraging groups.

In Panama, Russell found that, when foraging, adult white-nosed coatis tend to stay on the outside of the groups, while the youngsters forage in the center. As they forage, the animals in the front and on the sides have first choice of food items, while the animals in the center and rear are stuck with whatever is left. As such, foraging in an elliptical format is not very efficient in reducing competition, but it may reduce predation on the youngsters.[11] However, Hirsch found that juvenile brown-nosed coatis were usually at or near the front of the group, and subadults were at the back. Hirsch suggested that the spatial formation of foraging groups had little to do with predation but reflected competition and social interactions.[6,20] Whether these differences in spatial structure of the groups reflect differences between brown-nosed and white-nosed coatis, differences in predation risk, differences in methodology, or some other factor is not known. I suspect that the speed of the foraging group may also be a factor; if a group is foraging quickly, many items in the leaf litter will be missed by the leading animals.[21] This would mean that coatis foraging toward the back of the group still have plenty of food items available to them.

Individual coatis in large groups spend less time watching for predators than do solitary coatis or those in small groups and may be able to forage more efficiently.[1,11,20] Adults foraging in groups can share vigilance duties: some coatis can have their head down rooting through the leaf litter, while their troop mates scan for danger. However, unlike in some mongoose species, there are no individuals that appear to take on the role of sentinel.[22] In a detailed study of vigilance and group structure in Argentina,[20] coatis on the periphery of groups were more vigilant (defined as scanning their surroundings) than those in the center, as were those that were more spread out while foraging. The age of the youngsters also mattered: coatis were more vigilant when the youngsters were less than six months of age. Hirsch also found that coatis detected most new food sources by smell, not sight, so vigilance while foraging was mainly for detection of predators rather than looking for food.[21] In other words, Hirsch's results indicated that spatial position while foraging reflected competition, and vigilance levels reflected each individual's perception of predation risk to itself and to the youngsters. It should be noted here that coati vigilance may be more related to hearing than sight. According to Hirsch, *"I think it is highly likely that coati vigilance is mostly to reduce noise, and listen for predators and/or look for movement. The ability of a coati to spot a non-moving sit-and-wait predator is pretty lousy."*[23]

Male Sociality

I also put radios on 13 male coatis – five in Garden Canyon, seven in Huachuca Canyon and one in Ramsey Canyon. Most male coatis did not survive long, with almost all being killed by predators within a few months of being captured. I was only able to monitor four that lived for a year or more, with one (M77) being monitored for almost three years before the batteries on his second radio collar failed. Based on weight and tooth wear, I estimated that he was almost three years old when I first captured him in May of 1996, making him almost six years old in January of 1999, when I lost track of him.

One other male that I monitored for a year (M42) lived along the crest of the Huachucas in upper Huachuca Canyon. He was extremely shy, and I had only a few glimpses of him during the entire time he was radiocollared. He was killed by a mountain lion 13 months after he was radiocollared. The third male (M187) was captured with Troop 5, which I assumed to be his natal troop (see above). He was monitored for 12 months, until he was killed by a mountain lion. One male in Ramsey Canyon (M86) was monitored for 13 months. He was first observed on March 28th, 1996 (although he was well known to the folks at the Ramsey Canyon Preserve before that), and he was recognizable from his large size and numerous scars. He was monitored opportunistically until he was radiocollared in December 1996. His signal disappeared in April 1997, and no sign was ever seen of him again.

When Kaufmann first described coati societies in Panama, he observed that adult males were always solitary except during the mating season, when one male was allowed to join each troop.[3] The early studies in Arizona seemed to confirm this,[24,25] and I seldom observed adult males with the troops outside of the mating season. Later work in Panama and at other sites has revealed that this is not always the case. Some troops have an adult male or two that accompanies them outside of the mating season. In Panama, Gompper and Krinsley observed adult males apparently delaying dispersal from the troops and observed previously solo males accompanying troops on occasion. In addition, they saw them associating with subadults, and also observed instances of males traveling together.[26] In Guatemala, Susan Booth-Binczik observed males regularly accompanying troops, but they were usually on the periphery of the groups, and did not stay with them for very long.[5] In El Tepozteco, Mexico, two adult males were observed frequently as part of a group under study.[27] Gilbert, working in the southern end of the Huachucas, often observed several males with the troop he was watching. He was feeding the animals,[28] so it's not clear if they were part of the troop or in the area for free handouts. In a long-term study in the Chiricahua Mountains of Arizona, McColgin did not find males associating with any troops.[10] In Costa Rica, Joel Sáenz found that troops usually had a male associated with them year-round.[29] However, Caso, studying coatis in Tamaulipas, observed males with troops only during the mating season.[30]

Among brown-nosed coatis in Brazil and Argentina, several studies have found an adult male to be part of the troop, perhaps more integrated than is usually seen among white-nosed coatis. In Argentina, adult males were observed with troops 85% of the time outside of the mating season. Each troop contained only one male, and they were with the troop for several continuous months at a time. They groomed and shared nests with troop members. In most cases, the troops that the males stayed in were not their natal troops.[31] However, another study in Brazil found that males were solitary except during the mating season.[32] It appears, however, that most adult male brown-nosed coatis are solitary, even if some are allowed access into troops. Limited studies of mountain coatis have not found males to be part of groups.[7] So there appears to be some flexibility in the sociality of males, although it has yet to be determined what factors determine when and if a male is allowed into a troop.

Hirsch and Gompper suggested that there were species-level differences in male coati sociality, with brown-nosed coatis more likely to have males as members of a troop.[33] It is notable, however, that all or most populations in which males regularly associate with troops live in very rich habitats and/or have access to human food,[5,6,27,29] indicating that male sociality may be related to food availability. Perhaps when food is abundant, females are less aggressive toward adult males. This might also explain the differences in male sociality when Kaufmann observed white-nosed coatis on Barro Colorado Island (no males in troops outside of the mating season), compared with when Gompper observed the same population almost 40 years later (some males in troops year-round). Survival was quite low during Kaufmann's study, indicating that food resources may have been limited. The number of adult males in an area might also affect how likely they are to associate with a troop; populations where male associated with troops appeared to have relatively higher ratio of males to females.

Of the male coatis that were radiocollared during my study, only one was commonly seen in the company of a troop. M86 was observed with Troop 1 in Ramsey Canyon in 4 of 27 observations outside of the mating season. Two other males were observed with troops: M14 was observed with Troop 3 once and Troop 6 once, and M187 was observed with Troop 5 and Troop 3 once. It wasn't unusual to locate males within 100 m of troops, and I occasionally observed unmarked males with troops, but I never observed adult males to be an integral part of a troop, as has been observed in Brazil and Argentina.[31] I observed two males together on several occasions (not counting a fight between M42 and M77). They were usually traveling together with no obvious sign of an agonistic encounter. But these associations were short-lived, and subsequent observations found the males on their own. Male coatis in my study area fought frequently, both during the mating season and occasionally during the rest of the year. Interestingly enough, in captivity adult males can sometimes co-exist peacefully with both males and females, especially if they have been raised together.[34,35]

Male raccoons appear to have a different social arrangement. Several studies have documented unrelated males forming small coalitions, where several individuals will associate and defend a common territory.[36,37] These appear to be simple arrangements, rather than complex bond formation as seen in female coatis.

Learning to Be Social

The winds shift in mid- to late September, and the Huachucas begin to dry out. The afternoon thunderstorms all but disappear as the wind direction switches to a westerly flow from the Pacific, instead of the southerly flow from Mexico. Except for the occasional hurricane from the west coast of Mexico (which can bring very heavy rainfall and cause substantial flooding), most precipitation comes from storms that generate in the northern Pacific and find their way down the California coast. The amount of rainfall that reaches southern Arizona is strongly affected by whether the Pacific Ocean is under the influence of an El Niño or La Niña pattern. The El Niño-southern oscillation (ENSO) is a climate pattern that occurs in the tropical Pacific at roughly 2-7-year intervals. An El Niño episode, resulting from warmer ocean temperatures in the western Pacific, usually brings increased winter rainfall to the southwestern U.S. and northwestern Mexico. La Niña has just the opposite effect, with decreased winter rains. These weather patterns also affect the monsoon rains in Arizona, but to a lesser extent, and affect fruit production in areas of tropical Central America, impacting food availability for coatis and other fruit eaters.[38]

At this point the youngsters are weaned and finding their own food. The next year and a half, perhaps even longer, young coatis must learn what to eat, how to detect and avoid predators, how to interact with other coatis, and how to locate such critical resources such as food, water, and safe bedding sites. Observations of captive coatis reveal how critical this period is; coatis raised in captivity exhibit markedly different behavior patterns, including being less vocal, having a smaller behavioral repertoire, and being more tolerant of other coatis.[28,34] Learning proper social behavior takes time: coatis mature slowly for their body size, achieving physical and sexual maturity at least one year later than their raccoon cousins.[39,40] This social learning is also reflected in their brain size. Coatis have significantly larger frontal cortex regions than raccoons or kinkajous, with female coatis having a significantly larger frontal cortex than males.[41] Among some species of primates, a larger prefrontal cortex, a region of the frontal cortex, is associated with larger group size, and indeed, the size of the prefrontal cortex may place constraints on maximum group size.[42]

Few studies of coatis to date have identified any real dominance hierarchy among individuals, but there does appear to be a hierarchy based on age and gender. Among males, older males may be given some deference, but among the members of troops, coatis under nine months of age seem to have access to whatever they want. They quickly learn that their mothers and other adult females in the troop will back them up and defend them against threats,[19,43] so they become very bold about approaching subadults and even adult males to steal food. Several youngsters may even gang up on a male or subadult and force them from the area.[28,44,45] As they approach their first mating season, however, their preferential status starts to change and they slide all the way to the bottom of the hierarchy.

When the troop members come back together in the fall, subadults are at the bottom of the pecking order and are picked on by everyone, including the new youngsters. Their mothers no longer come to their defense. This odd social order was best described by Hirsch for brown-nosed coatis[44] but was also observed in white-nosed coatis.[3,28,45] It differs from raccoons, which appear to have age-related hierarchies when aggregating around food sources.[46] It was too difficult for me to see the coatis well enough to quantify these interactions, but nothing I observed contradicted them. Most notably I observed that, at least when resting, the subadults were the most vigilant of the group and usually the first to send out the alarm. Luckily for someone trying to sneak up on the group, the subadults were often ignored.

A recent, more detailed, study of dominance hierarchies in white-nosed coatis found a moderately linear hierarchy, based somewhat on age and gender (older females at the top, adult males at the bottom), and identified coalition size as an important factor in dominance interactions.[27] In other words, the larger the adult female-juvenile subgroup, the higher their ranking. This could also be a somewhat circular argument – higher ranking females may have higher survival of their offspring, leading to larger coalition size. The study was somewhat limited; it focused on one troop (comprising five coalitions) for only eight months, and on only a few behaviors, ignoring the role of vocalizations and subtle facial expressions. Nevertheless, it did identify that coatis maintain long-term social relationships that may serve to reduce the amount of strife within the group.[27]

Notes and References

1. Burger J, Gochfeld M (1992) Effect of group size on vigilance in the coati, *Nasua narica* in Costa Rica. Animal Behaviour 44:1053–1057

2. Gompper ME (1996) Sociality and asociality in white-nosed coatis (*Nasua narica*): foraging costs and benefits. Behavioral ecology 7:254–263

3. Kaufmann JH (1962) Ecology and social behavior of the coati, *Nasua narica*, on Barro Colorado Island, Panama. University of California Publications in Zoology 60:95–222

4. Valenzuela D (1999) Efectos de la estacionalidad ambiental en la densidad, la conducta de agrupamiento y el tamaño del área de actividad del coatí (*Nasua narica*) en selvas tropicales caducifolias. Ph.D. Disertación, Instituto de Ecología, UNAM, México, D.F.

5. Booth-Binczik SD (2001) Ecology of coati social behavior in Tikal National Park, Guatemala. Ph.D. Dissertation, University of Florida, Gainesville

6. Hirsch BT (2007) Within-group spatial position in ring-tailed coatis (*Nasua nasua*): balancing predation, feeding success, and social competition. Ph.D. Dissertation, Stony Brook University, Stony Brook, NY

7. Rodriguez-Bolanos A, Sanchez P, Cadena A (2003) Patterns of activity and home range of mountain coati, *Nasuella olivacea*. Small Carnivore Conservation 29:16–19

8. Gompper ME, Gittleman JL, Wayne RK (1998) Dispersal, philopatry, and genetic relatedness in a social carnivore: comparing males and females. Molecular Ecology 7:157–163

9. Hass CC (2002) Home-range dynamics of white-nosed coatis in southeastern Arizona. Journal of Mammalogy 83:934–946

10. McColgin ME, Koprowski JL, Waser PM (2018) White-nosed coatis in Arizona: tropical carnivores in a temperate environment. Journal of Mammalogy 99:64–74

11. Russell JK (1983) Altruism in coati bands: nepotism or reciprocity? *In*: Wasser SK (ed) Social behavior of female vertebrates. Academic Press, New York, NY, pp 263–290

12. Gompper ME, Gittleman JL, Wayne RK (1997) Genetic relatedness, coalitions and social behaviour of white-nosed coatis, *Nasua narica*. Animal Behaviour 53:781–797

13. Moss CJ (2012) Elephant memories: thirteen years in the life of an elephant family. University of Chicago Press, Chicago, IL

14. Avilés L, Fletcher JA, Cutter AD (2004) The kin composition of social groups: trading group size for degree of altruism. The American Naturalist 164:132–144

15. A "location" being a record in which I was able to determine the spatial coordinates of the radiocollared coati. I excluded records in which I could determine, say, that a radiocollared coati was somewhere in the canyon, but could not determine more precisely where it was.

16. Association percentages calculated from a Sorensen index, considering only the amount of time the pairs were both wearing active radio collars.

17. "Pairs" here implies no relationship or bond, just a unit for analysis.

18. Romero T, Aureli F (2007) Spatial association and social behaviour in zoo-living female ring-tailed coatis (*Nasua nasua*). Behaviour 144:179–193

19. Hirsch BT, Stanton MA, Maldonado JE (2012) Kinship shapes affiliative social networks but not aggression in ring-tailed coatis. PLoS ONE 7:e37301

20. Di Blanco Y, Hirsch BT (2006) Determinants of vigilance behavior in the ring-tailed coati (*Nasua nasua*): the importance of within-group spatial position. Behavioral Ecology and Sociobiology 61:173–182

21. Hirsch B (2010) Trade-off between travel speed and olfactory food detection in ring-tailed coatis (*Nasua nasua*). Ethology 116:671–679

22. Rood JP (1986) Ecology and social evolution of mongooses. *In*: Rubenstein DI (ed) Ecological aspects of social evolution. Birds and mammals. Princeton University Press, Princeton, NJ, pp 131–152

23. Ben Hirsch, personal communication.

24. Risser SC Jr (1963) A study of the coati mundi *Nasua narica* in southern Arizona. Master's Thesis, University of Arizona, Tucson

25. Wallmo OC, Gallizioli S (1954) Status of the coati in Arizona. Journal of Mammalogy 35:48–54

26. Gompper ME, Krinsley JS (1992) Variation in social behavior of adult male coatis (*Nasua narica*) in Panama. Biotropica 24:216–219

27. de la O C, Furtbauer I, King AJ, Valenzuela-Galván D (2019) A resident-nepotistic-tolerant dominance style in wild white-nosed coatis (*Nasua narica*)? Behaviour 1–42

28. Gilbert B (1973) Chulo. A year among the coatimundis. Alfred A. Knopf, New York, NY

29. Sáenz JM (1994) Ecología del pizote (*Nasua narica*) y su papel como dispersador de semillas en el bosque seco tropical, Costa Rica. Master's Thesis, Universidad National, Heredia, Costa Rica

30. Caso A (1994) Home range and habitat use of three neotropical carnivores in northeast Mexico (*Felis pardalis, Felis yagouaroundi, Nasua narica*). Master's Thesis, Texas A&M, Kingsville

31. Hirsch BT (2011) Long-term adult male sociality in ring-tailed coatis. Mammalia 75:301–304

32. de Barros D, de Cassia Frenedozo R (2010) Uso do habitat, estrutura social e aspectos básicos da etologia de um grupo de quatis (*Nasua nasua* Linnaeus, 1766) (Carnivora: Procyonidae) em uma área de Mata Atlântica, São Paulo, Brasil. Biotemas 23:175–180

33. Hirsch B, Gompper ME (2017) Causes and consequences of coati sociality. *In:* Macdonald DW, Newman C, Harrington, L.A. (eds) Biology and conservation of Musteloids. Oxford University Press, Oxford, UK, pp 515–526

34. Kaufmann JH, Kaufmann A (1963) Some comments on the relationship between field and laboratory studies of behaviour, with special reference to coatis. Animal Behaviour 11:464–469

35. Stan and Linda Rolinski, personal communication.

36. Gehrt SD, Gergits WF, Fritzell EK (2008) Behavioral and genetic aspects of male social groups in raccoons. Journal of Mammalogy 89:1473–1480

37. Hirsch BT, Prange S, Hauver SA, Gehrt SD (2013) Genetic relatedness does not predict raccoon social network structure. Animal Behaviour 85:463–470

38. Wright SJ, Carrasco C, Calderon O, Paton S (1999) The El Niño southern oscillation, variable fruit production, and famine in a tropical forest. Ecology 80:1632–1647

39. Gehrt SD, Fritzell EK (1998) Duration of familial bonds and dispersal patterns for raccoons in south Texas. Journal of Mammalogy 79:859–872

40. Zeveloff SI (2002) Raccoons: a natural history. Smithsonian Books, Washington, D.C.

41. Arsznov BM, Sakai ST (2013) The procyonid social club: comparison of brain volumes in the coatimundi (*Nasua nasua, N. narica*), kinkajou (*Potos flavus*), and raccoon (*Procyon lotor*). Brain, Behavior and Evolution 82:129–145

42. Dunbar RIM (1992) Neocortex size as a constraint on group size in primates. Journal of Human Evolution 22:469–493

43. Romero T, Aureli F (2008) Reciprocity of support in coatis (*Nasua nasua*). Journal of Comparative Psychology 112:19–25

44. Hirsch BT (2007) Spoiled brats: is extreme juvenile agonism in ring-tailed coatis (*Nasua nasua*) dominance or tolerated aggression? Ethology 113:446–456

45. Russell JK (1981) Exclusion of adult male coatis from social groups and protection from predation. Journal of Mammalogy 62:201–206

46. Hauver S, Hirsch BT, Prange S, Dubach J, Gehrt SD (2013) Age, but not sex or genetic relatedness, shapes raccoon dominance patterns. Ethology 119:769–778

7

Benefits of Group Living

Why coatis live in groups has been the subject of much study and debate. In the initial studies of coatis on Barro Colorado Island in Panama, Russell suggested that female coatis live in groups to protect the young from predation (cannibalism) by the larger males.[1] Russell observed several instances of adult males attacking and, in one case, killing juveniles. However, no other studies have reported similar findings. Gompper, in a later study of foraging behavior in the coatis at the same location, found that solitary males foraged more efficiently (obtained more food items per unit time) than social females when eating fruits, and solitary females foraged more efficiently than social females when eating invertebrates. He also found that adult males could supplant solitary females and small groups from fruiting trees, although larger groups could fend off males, and he suggested that coati sociality evolved as a way for females to maintain access to preferred foods, particularly fruits.[2] Similar results have been found for some species of primates.[3]

Foraging Benefits

In contrast, Booth-Binczik found that, although females foraged more efficiently when alone, there were no differences in foraging rates between males that were solo or those that were in groups. Nor did males usurp fruits from females.[4] In Arizona, all of the fruits available to coatis are small and very abundant for short periods. I never saw coatis defending fruits or even fruiting trees or bushes. So, the foraging benefits of sociality are not clear, and may vary from site to site depending on the availability of large fruit and fruit patches. A tree with a thousand juniper berries might not be worth defending, but one with a few ripe papayas might be. In addition, mountain coatis, with a diet of almost exclusively insects and other invertebrates[5] are also social, living in groups of up to 50 individuals, so access to fruit would not explain sociality in those species.

Everyone who has spent any amount of time watching coatis forage has seen that they do not like to share their food. Tasty morsels are vigorously defended. Nor do they cooperate in hunting, like wolves. After all, what would be the point of ganging up on a bug? John Kaufmann suggested, based on casual observation, that foraging success might be higher in groups for items like lizards, where a lizard that escapes capture by one coati might fall prey to another nearby.[6] No data have been collected to test his hypothesis.

Coalitions

The formation of coalitions seems to be a defining aspect of coati sociality. Adult females gang up on adult males and each other, youngsters (backed up by adult females) gang up on males and other adult females, and everyone gangs up on subadults. According to studies on brown-nosed coatis, coalitional support appears to be as complex as it is in many species of primates and may involve a form of score-keeping.[7] Ben Hirsch observed an extraordinary amount of aggression among brown-nosed coatis at Iguazu National Park, Argentina, and suggested that constant aggression toward subadults may encourage their leaving as a group and forming their own troop.[8] No other studies have documented as much aggression toward subadults as Hirsch did, nor have they documented subadults leaving as a group. However, even with lower levels of aggression, Kaufmann and Gilbert noticed that subadult white-nosed coatis were more likely to stray, and Jim Russell observed that recent migrants into a troop were usually two-year-olds.[6,9,10] If so, given the costs and benefits of sociality (see below), we might expect that aggression toward subadults would increase with increasing group size.

I saw a couple of instances of females ganging up on adult males. On April 8, 1997, I saw the members of Troop 4 resting on a ledge of a cliff in Huachuca Canyon. M37 rested on a nearby ledge about 23 m (75 ft) away.

After an hour or so, several adult females from the troop got up and walked along the ledge toward M37. He stood and backed up along the ledge, facing the approaching females. He continued backing up as they approached. When they were almost nose-to-nose, he ran out of room to back up and fell off the ledge, landing in a prickly pear cactus about 10 m (30 ft) below. I could hear him scream from my perch across the canyon. As the females casually walked back to the rest of the group, he clambered up to another ledge and started pulling out the cactus spines with his teeth. The females regrouped and headed off in the other direction. This observation occurred during the mating season, but it was not clear why the females approached M37.

On December 10, 1997, I was watching Troop 1 foraging for madrone berries on the Ramsey Canyon Preserve. Most of the coatis foraged up in the treetops, but a few were on the ground, picking up morsels dropped by animals above. Suddenly, a large male rushed into the group and attacked a coati that was foraging on the ground. It happened so fast that I did not see which animal was attacked but within seconds at least half a dozen adult females rushed over and attacked the male. For several seconds, all I could see was a ball of squealing and snarling coatis but then the group broke up, and the male chased two females while the rest of the group fled in the other direction. The first female (unmarked) ran up a small tree, quickly followed by the male. She backed up on a small branch, and he followed as they squealed and bit at each other's noses. She finally lost footing with her back feet and dangled for a few moments by her front feet before dropping about two m (six ft) to the ground and running off. The male descended the tree by climbing down the trunk and ran off after the second female, who had been perched in a nearby tree, but she quickly ran after the rapidly retreating troop. The male started to head away from the group and down the canyon, first pausing to stand on a rock and glance around. I saw blood dripping from his nose. This interaction continues to puzzle me, although when I described it to John Kaufmann, he mentioned that he had seen similar interactions in Panama. Was this the type of interaction that Jim Russell interpreted as attempted predation on juveniles by adult males?

Protection from Predators

During the earlier studies on Barro Colorado, few large predators existed on the island. Smaller predators, such as ocelot, tayra, and boa constrictors were present, but none were seen killing coatis during any of the studies, thus it was not possible to study coati-predator interactions. Boas and ocelots have killed coatis in other areas of Latin America, however. In one observed attack of a coati by a boa constrictor in Costa Rica, several group members tried to aid the captured coati by biting and clawing at the snake. They were not successful, and the coati succumbed to the constrictor.[11] This observation illustrates both that coatis are vulnerable to boas, and that they will cooperate in defense. A subsequent study of mountain lion and ocelot diets on Barro Colorado Island found remains of coatis in the scats of both,[12] nonetheless predation risk appeared to be much lower than in some other study areas.

In the Huachucas, I was able to determine cause of death for 25 of the 38 coatis that were radiocollared. Of those 25, 19 were killed by large predators. By saying they were killed by predators, I do not mean to imply that I observed them being killed. The radio collars on the coatis were equipped with mortality sensors that doubled the pulse rate of the "beeps" if the collar did not move for eight hours. This enabled me to find carcasses within a few days (sometimes a few hours) after the collar stopped moving. Different predators kill and consume their prey in different ways, and we soon learned to recognize the characteristics signs of mountain lion and black bear predation. For example, mountain lions typically plucked the hair from the carcass then consumed almost all of it, leaving behind the nose and muzzle, last six inches of the tail, and sometimes a paw or two. They typically crushed the back of the skull and consumed the brain. Usually the remains were not buried, as they often are for larger prey. Black bears, on the other hand, tended to peel a coati from back to front, much like when you pull a t-shirt off over your head (starting at the bottom hem, not the neck band). They would pull the hide back to expose everything except the head and front legs, and then eat the meat off the bones.

Eighteen of the radiocollared coatis appeared to have been killed by mountain lions, and one by a black bear. We also found other carcasses of coatis that were not radiocollared and heard of instances of other mortalities. Of those, five were killed by mountain lions (one juvenile, four adults), one by a black bear (adult), one by a spotted owl (observed - juvenile), four by dogs (three observed - one adult, two juvenile), four were killed by cars, and one by electrocution at an electrical substation. So, in all, of 41 known mortalities, 73% were due to predation, and 77% of the predation was by mountain lions.

In Mexico, Valenzuela also documented predation on coatis by big cats – both mountain lions and jaguars. We collaborated to examine mortality rates in coatis, and to see if living in groups conferred any advantage in the face of large predators. We found that in both populations, solitary coatis (males and females away from the troop to give birth) were much more vulnerable to predation than were coatis in troops. Overall, predation rates were twice as high on males as on females. Predation risk was almost five times higher for females when they were solitary during the nesting period in Arizona, and almost nine times higher during the same period in Mexico, then when they were with their troops.[13] Most of the females killed by predators during the nesting season were killed just before or just after they gave or would have given birth. Overall, adult males had a 41% chance of being killed by predators each year (which might explain why there were so few old male coatis), in contrast to a 12% chance each year for females in troops.

In Arizona, but not Mexico (due to insufficient data at the latter site), predation rates were higher on smaller groups than on larger groups. Predation rates were much higher if there were 10 or fewer animals in a troop than if there were more than 10.[14] Likewise, predation rates were higher if there were fewer than four adult females. In Guatemala, where jaguars prowled the jungle outside of the Mayan ruins at Tikal, Booth-Binczik also found that predation rates were higher for females that were solitary or in small groups.[4,15] And Hirsch, studying brown-nosed coatis near Iguazu Falls, Argentina, noted that mortality on adult females was highest during the nesting season when the females were solitary, although he was unable to document the cause of the mortalities.[16] For the researcher, small groups of coatis appear to be much more easily spooked and difficult to approach.[17,18] .

After recovering several coati carcasses that had been killed by mountain lions, it became clear to me that mountain lions were of great significance to coatis. But were coatis particularly important to mountain lions? Mountain lions are generally considered deer specialists, so I was curious how much of the local lion's diet consisted of coatis. I started collecting mountain lion scats (feces) in 1997 to see what the cats were eating. That winter I also noticed that there was a lot of bobcat activity (evidenced by large numbers of scats) in the same area as a couple of the coati troops, so

I started collecting those scats as well. Analysis of 125 very stinky cat scats revealed that coatis were quite important to mountain lions – between 1997 and 2002, they were the second most common item on the big cat's menu, after deer (mule and white-tailed). Bobcats, too, were chomping on the occasional coati, but the coati remains in bobcat scats were juvenile coatis, and they were not particularly important, making up less than 5% of the diet.[22]

That coatis might be part of a mountain lion's menu was at first met with skepticism by some local mountain lion researchers (accustomed to prey studies outside of coati range), but most studies of mountain lion diets from the tropics have found that coatis are a common dietary item.[23-30] One study in Costa Rica found that coatis were the number one item in mountain lion scats.[12] They are also commonly found in the diets of jaguars,[23-25,27,28,31-35] and they show up occasionally in the diets of margays and ocelots.[36,37] Other known predators throughout coati range include hawks, and eagles.[9,38-39] At one site in Costa Rica, nestling coatis were vulnerable to predation from capuchin monkeys.[19-21,40] The solitary coati mothers were often unable to protect their nests from marauding monkeys, especially if several monkeys were involved.

I was fortunate enough to see a couple of interactions between coatis and their potential predators. On April 7, 1997, I located F33 and F69 from Troop 3 in the cliffs of middle Garden Canyon. The two females and one juvenile (10 months old) were perched in a couple of pinyon pines high up on a cliff and appeared agitated at something below them. Although I could not hear them above the noise of the stream, I could see that they were rapidly opening and closing their mouths, as if they were issuing alarm barks, and rapidly swishing their tails back and forth. They also moved back and forth on the tree branches, as if tracking the movements of something below them. As I looked to see what they were watching, I saw a mountain lion pacing back and forth under the trees. I only got a brief glimpse of its tawny body as it came into view and quickly faded away, but watching the coatis gave me a fairly good idea where it was. They remained agitated for more than an hour, then finally huddled together, and seemed to be resting.

I left to find the rest of the troop, and found them asleep, high up in a ponderosa pine in Sawmill Canyon, nearly a mile away as the crow flies. As I stood under the tree taking notes, elated that I was able to get so close, they awoke from their nap and began stretching, urinating, and defecating. Suddenly it was raining coati pee and poop, elation turned to disgust, and I made a quick exit. It was late afternoon by the time I returned to the cliff where F33 and F69 were, and I saw they were still huddled together in the top of the tree, more than six hours after I first observed them. I searched the cliff a few days later (the females had rejoined the rest of the troop by then), and I found a fresh mountain lion scat with coati hair and claws in it, although none of the marked animals were missing.

On October 14, 1999, I found Troop 3 in the cliffs in Garden Canyon, not far from the above sighting. They were resting on ledges and in a couple of small trees, and although I had signals from three radiocollared females, I could only see a couple of unmarked animals. I found a comfortable spot to sit on the opposite hillside and waited for the animals to start foraging so I could get a count. After an hour of hiding in the shade, my butt was getting sore, I was getting chilled, and the coatis showed no sign of moving. Just then a dark shadow flew over my head in the direction of the coatis. The golden eagle passed right over the sleepy coatis, banked, and slowly passed over them again. The trees on the cliff started raining coatis, and I could hear an assortment of alarm calls and contact calls as they huddled under the small trees. The eagle made one more unhurried pass, almost hovering, with its talons just brushing the tops of the trees, before it casually flew out of sight down the canyon. After the eagle disappeared, chaos ensued as coatis ran every direction in a panic. Finally, after several minutes of running this way and that, the group coalesced and made an orderly retreat up the canyon. Luckily for me, they progressed single file along a ledge, and I was able to get a good count: 39 animals total including 26 youngsters. A large, nearly black, adult male was with the group, although he did not stay with them as they headed up the canyon. I found no evidence of successful predation by golden eagles, but other researchers have.[39]

In late 1996, I found a dead 5-month-old coati lying next to the dirt road going up Garden Canyon. I collected the coati kit, and upon examination, discovered large puncture wounds on top of the skull, a severed trachea, and a couple sets of puncture wounds on the ribcage, plus some broken ribs. The spacing of the puncture wounds and extent of the injuries were consistent with mountain lion predation.[17,41] But what caused the lion to drop the baby coati? Perhaps the troop came to the baby's defense?

Michael Sutor, during the filming of *Nasenbären*, just missed witnessing one of the radiocollared females being killed by a mountain lion. I had left my telemetry gear with him while I took a couple of days off from field work during January 1999. He found Troop 3 in a side canyon of Garden

Canyon one morning. He filmed the coatis as they stayed in the trees, seemingly tolerating his presence, for an entire day. They were still in the trees the next day, and as he scanned through the frequencies on the telemetry receiver, he found that the signal from F33 (the same female from the previous mountain lion story) was in mortality mode. He located the source of the signal and found the radio collar and a pile of bloody hair. He left the canyon and found a phone to call me. I caught up with him a couple of hours later. The coatis had left the trees and they crossed most of their home range in a hurry, with Michael tagging along – 4 km (2.5 mi) from where the troop had started that morning.

Michael graciously left me some extra footage from his filming, including the footage of the coatis that he had filmed that day. In the tape, they spent most of their time looking at the ground at the base of the tree. They were quiet and seemed reluctant to move far or to descend to forage (they were up in a couple of oak trees that had no acorns at the time). This is very unusual behavior for an animal that spends most of its day foraging. Although they started to move into other adjoining trees by late afternoon, they did not come down to the ground. Based on what I saw in the tape, and what I observed on the ground (the kill site was adjacent to the tree he filmed the coatis in), I think F33 must have been killed just a few hours before Michael got there. And I think that, based on the reluctance of the coatis to descend from the tree, the mountain lion was still in the area when he was filming.

While group living appears to provide significant survival benefits to adult female coatis, females in groups were still vulnerable. Young coatis also benefit from being in a group. As mentioned in the last chapter, having similar-aged playmates may be beneficial in the development of social behavior through play. They benefit from being able to form coalitions for protection and to access valued resources. Coatis tend to sleep in a pile, so there may be some thermal benefit to being in a group. The adult females take care of all the youngsters in a troop: they occasionally nurse and groom each other's young, and all will come to the defense of any youngster in distress. The youngsters also benefit through cultural transmission of knowledge of critical resources like the locations of water, nesting areas, and when and where particular food sources are available.[42] They also might learn the best escape routes and hiding places when confronted with a predator. The importance of cultural transmission is still a

90

seriously understudied aspect of coati behavior, but has been shown to be critically important for other social species.[43]

Protection from Parasites

Coatis spend a lot of time grooming, both themselves and each other. Allogrooming (meaning, grooming "others") may be important for social bonding as well as for removing parasites. In Panama, Gompper noted that females in groups had fewer ticks but more mites than did males, and he concluded that allogrooming was beneficial in controlling ticks.[44,45] In Arizona, none of the coatis I handled had any visible ticks, but they did have fleas, lice, and mites. Adult females had slightly more of each than did adult males, juveniles had only a few fleas and no other parasites, but subadults had the highest levels of all three parasites. In this case, allogrooming may have been beneficial to the juveniles, who received the most grooming from the adult females. But being in a group may have been detrimental to the subadults, who received little grooming from the adult females[46] yet had the misfortune of being in close enough contact to "share" the groups' parasites. The parasite levels I saw in the coatis were low and appeared to have little impact on the health of the animals. I did observe scabies infections (sarcoptic mange) caused by mites in skunks[47] and opossums during the study, but no such infections in the coatis. So at least for some animals in the troop, including adult males lucky enough to associate with a troop, one of the benefits of group living was fewer external parasites.

Social Behavior in Other Carnivorans

The trade-off between reduced predation risk and increased foraging benefits has been an ongoing debate regarding sociality in a large number of vertebrates.[4,49,50] Some mongooses, including meerkats, exhibit some superficial similarities to coatis. Some 36 species of mongooses are scattered throughout Africa and parts of Asia, as well as introduced to some Caribbean islands. Some species are solitary, and others live in groups. The group-living mongooses are generally small, diurnal, and insectivorous. They live in packs of up to 30 animals, and all cooperate to raise the young by bringing them food (insects) and scanning for predators. Juveniles are allowed to take food from older animals and are supported by adults in conflicts. In areas with lots of predators, mortality is lower in larger groups. In areas with few predators, however, mortality is higher in larger groups, presumably due to competition for food.[49] However, in most of the social mongooses, mating is limited to a few high-ranking animals, and subadults that have not dispersed yet act as "helpers," by babysitting, acting as sentries around the den, and providing food. One or more adult males live in the pack. The social mongooses are also usually found in open habitats, unlike the forested habitats where coatis live.[49,51,52]

This type of cooperative breeding, which is also found in some other mammal and bird species, has been of great interest to biologists who seek to explain why these helpers would forgo their own reproduction to care for someone else's offspring. A variety of explanations have been put forth, including reciprocal altruism (in essence, I'll help you now if you help me later); kin selection (by helping my brothers and sisters I'm also increasing my own genetic fitness); and an incidental form of assistance being provided by offspring who are delaying reproduction until suitable territories become available.[51,53–55]

Cooperative Breeding Systems

The circumstances surrounding the details of cooperative breeding systems and the factors that led to them are the subjects of several books, so I won't review all of the details here, except to say that cooperative breeding exists on a continuum. According to Tim Clutton-Brock,[51] cooperative breeders can be placed into four broad categories. There are group breeders, in which multiple females live and breed in the same social groups. They may gang up to defend resources, help detect predators, but there is little or no alloparental (meaning parenting "others") care. Kangaroos, some bats, ungulates, some primates and some carnivorans fall into this group. Then there are the communal breeders, in which multiple breeding females share in the care of the young in the group. Not all adult females breed, and the parents may be assisted by non-breeding females or males. Examples here include African lions, banded mongooses, spotted hyenas, some bats, and some primates. Facultative cooperative breeders come next, distinguished by the presence of non-breeding helpers, although they can successfully rear young without helpers. Examples include silver-backed jackals and European foxes, marmosets, and tamarins (the latter two being tiny primates). I would include wolves here also. Finally, there are specialized cooperative breeders, which must have helpers to successfully rear young. These helpers are not breeders, rather they delay reproduction and stay in their natal territory to aid the breeding pair. Meerkats fall into this last category.

Coatis might be classified as communal breeders according to this system.[49,56] All or nearly all adult females breed, and although they give birth by themselves, they rear their young in a communal group. Even females that might have lost their litters before returning to the troop help in caring for the young. Adult females nurse and groom each other's kits and come to their defense against predators and in conflicts with other coatis. Their vigilance protects all the youngsters in the group, not just their own. Much of troop life revolves around the youngsters. Although Russell documented the cooperative nature of coati groups,[10] they have been pretty much ignored in discussions of cooperative breeding. This type of communal care in mammals appears to be quite rare.[56]

Where coatis differ from other communal breeders is in the role of the subadults. Among most cooperative breeders, subadults provide much of the helping behavior toward the juveniles. Subadult coatis really seem to be outcasts in coati groups; they receive and give little cooperative care toward the youngsters or other members of the group. They do help scan for predators and will engage in play with the juveniles. However, as mentioned above, they often seem to be ignored when they alarm call, and otherwise seem to be waiting to mature. Male subadults are waiting to become mature enough to head out on their own. Female subadults are waiting to become old enough to breed and contribute their own offspring to the group. There were so few subadults alive during my study (usually fewer than three per troop), that I may have a biased perspective of the role they play in the group. Ben Hirsch reported an extraordinary amount of aggression directed toward subadults by all other age/sex classes, but he also had a large, rapidly growing population.[8] To date, most studies have focused on the adults (probably because they are easier to observe). More information is needed on the role of the subadults. Is their role simply to augment the group size, potentially reducing predation risk, while they finish their physical and social maturation? In this case, more aggression toward subadults might be expected as group size start to increase, and they become less critical to the group. As predation risk seems to vary in different areas, we might expect aggression toward subadults to be both a function of group size and predation risk.

Subadults may also be the dispersal units for coati populations. When the pregnant females leave the troops to give birth, the subadults (one- to two-year-olds) are left behind. Small groups of subadults (two to five individuals) have been observed outside their normal range during this period,[17,57] and Hirsch suggested that subadults left their troops to form new ones.[8] Although coatis are getting more attention from carnivore researchers, there is still much we don't know about the ecological factors that influence sociality.

Notes and References

1. Russell JK (1981) Exclusion of adult male coatis from social groups and protection from predation. Journal of Mammalogy 62:201–206

2. Gompper ME (1996) Sociality and asociality in white-nosed coatis (*Nasua narica*): foraging costs and benefits. Behavioral Ecology 7:254–263

3. Wrangham RW (1986) Ecology and social relationships in two species of chimpanzee. *In*: Ecological aspects of social evolution. Birds and mammals. Princeton University Press, Princeton, NJ, pp 352–378

4. Booth-Binczik SD (2001) Ecology of coati social behavior in Tikal National Park, Guatemala. Ph.D. Dissertation, University of Florida, Gainesville

5. Rodriguez-Bolanos A, Cadena A, Sanchez P (2000) Trophic characteristics in social groups of the mountain coati, *Nasuella olivacea* (Carnivora: Procyonidae). Small Carnivore Conservation 23:1–6

6. Kaufmann JH (1962) Ecology and social behavior of the coati, *Nasua narica*, on Barro Colorado Island, Panama. University of California Publications in Zoology 60:95–222

7. Romero T, Aureli F (2008) Reciprocity of support in coatis (*Nasua nasua*). Journal of Comparative Psychology 112:19–25

8. Hirsch BT (2007) Spoiled brats: is extreme juvenile agonism in ring-tailed coatis (*Nasua nasua*) dominance or tolerated aggression? Ethology 113:446–456

9. Gilbert B (1973) Chulo. A year among the coatimundis. Alfred A. Knopf, New York, NY

10. Russell JK (1983) Altruism in coati bands: nepotism or reciprocity? *In*: Wasser SK (ed) Social behavior of female vertebrates. Academic Press, New York, NY, pp 263–290

11. Janzen D (1970) Altruism by coatis in the face of predation by boa constrictor. Journal of Mammalogy 51:387–389

12. Bustamante A, Moreno R, Sáenz JC (2009) Predation of a coati (*Nasua narica*) by a puma (*Puma concolor*) in the southeast of the Osa Peninsula, Costa Rica. Acta Biologica Panamensis Vol 1:39–45

13. Predation risk is defined here as the annualized daily predation rate, calculated from the interval predation rate, and extrapolated for the whole year. See Hass and Valenzuela (2002) for calculations.

14. Hass CC, Valenzuela D (2002) Anti-predator benefits of group living in white-nosed coatis (*Nasua narica*). Behavioral Ecology and Sociobiology 51:570–578

15. Note that Booth-Binczik assumed predation if animals disappeared with no sign of illness or injury. No actual evidence of predation was found.

16. Hirsch BT (2007) Within-group spatial position in ring-tailed coatis (*Nasua nasua*): balancing predation, feeding success, and social competition. Ph.D. Dissertation, Stony Brook University, Stony Brook, NY

17. Personal observation.

18. Ben Hirsch, personal communication.

19. Fedigan LM (1990) Vertebrate predation in *Cebus capuchinus*: meat eating in a neotropical monkey. Folia Primatologica 54:196–205

20. Newcomer MW (1985) White-faced capuchin (*Cebus capucinus*) predation on a nestling coati (*Nasua narica*). Journal of Mammalogy 66:185–186

21. Perry S, Rose L (1994) Begging and transfer of coati meat by white-faced Capuchin monkeys, *Cebus capuchinus*. Primates 35:409–415

22. Hass CC (2009) Competition and coexistence in sympatric bobcats and pumas. Journal of Zoology 278:174–180

23. Facure KG, Giaretta AA (1996) Food habits of carnivores in a coastal Atlantic forest of southeastern Brazil. Mammalia 60:499–502

24. Jorgenson JP, Redford KH (1993) Humans and big cats as predators in the Neotropics. Symposium of the Zoological Society of London 65:367–390

25. Moreno RS, Kays RW, Samudio R (2006) Competitive release in diets of ocelot (*Leopardus pardalis*) and puma (*Puma concolor*) after jaguar (*Panthera onca*) decline. Journal of Mammalogy 87:808–816

26. Monroy-Vilchis O, Gomez Y, Janczur M, Urios V (2009) Food niche of *Puma concolor* in central Mexico. Wildlife Biology 15:97–105

27. Nuñez R, Miller B, Lindzey FG (2000) Food habits of jaguars and pumas in Jalisco, Mexico. Journal of Zoology 252:373–379

28. Ávila-Nájera DM, Palomares F, Chávez C, Tigar B, Mendoza GD (2018) Jaguar (*Panthera onca*) and puma (*Puma concolor*) diets in Quintana Roo, Mexico. Animal Biodiversity and Conservation 41.2:257–266

29. Jaimes RP, Caceres-Martinez CH, Acevedo AA, Arias-Alzate A, Gonzalez-Maya JF (2018) Food habits of puma (*Puma concolor*) in the Andean areas of Tamá National Natural Park and its buffer zone, Colombia. Therya 9:201–208

30. Soria-Díaz L, Fowler MS, Monroy-Vilchis O (2017) Top-down and bottom-up control on cougar and its prey in a central Mexican natural reserve. European Journal of Wildlife Research. https://doi.org/10.1007/s10344-017-1129-y

31. Hoogesteijn R, Hoogesteijn A, Mondolfi E (1993) Jaguar predation and conservation: cattle mortality caused by felines on three ranches in the Venezuelan Llanos. Symposium of the Zoological Society of London 65:391–407

32. Novack AJ, Main MB, Sunquist ME, Labisky RF (2005) Foraging ecology of jaguar (*Panthera onca*) and puma (*Puma concolor*) in hunted and non-hunted sites within the Maya Biosphere Reserve, Guatemala. Journal of Zoology 267:167–178

33. Rabinowitz AR, Nottingham BG Jr (1986) Ecology and behaviour of the jaguar (*Panthera onca*) in Belize, Central America. Journal of Zoology, London 210:149–159

34. Tewes ME, Schmidly DJ (1987) The neotropical felids: jaguar, ocelot, margay, and jaguarundi. *In*: Nowak M (ed) Wild furbearer management and conservation in North America. Ontario Ministry of Natural Resources, Toronto, pp 697–711

35. Hernández-SaintMartín AD, Rosas-Rosas OC, Palacio-Núñez J, Tarango-Arambula LA, Clemente-Sánchez F, Hoogesteijn AL (2015) Food habits of jaguar and puma in a protected area and adjacent fragmented landscape of Northeastern Mexico. Natural Areas Journal 35:308–317

36. Mondolfi E (1982) Notes on the biology and status of the small wild cats in Venezuela. *In*: Miller SD (ed) Cats of the world: biology, conservation, and management. National Wildlife Federation, Washington, D.C., pp 125–146

37. Wang E (2002) Diets of ocelots (*Leopardus pardalis*), margays (*L. wiedii*), and oncillas (*L. tigrinus*) in the Atlantic rainforest in southeast Brazil. Studies on Neotropical Fauna and Environment 37:207–212

38. Binczik GA (2006) Reproductive biology of a tropical procyonid, the white-nosed coati. Ph.D. Dissertation, University of Florida, Gainesville

39. Risser SC Jr (1963) A study of the coati mundi *Nasua narica* in southern Arizona. Master's Thesis, University of Arizona, Tucson

40. Sáenz JM (1994) Ecología del pizote (*Nasua narica*) y su papel como dispersador de semillas en el bosque seco tropical, Costa Rica. Master's Thesis, Universidad Nacional, Heredia, Costa Rica

41. Harley Shaw, personal communication.

42. See also Rodriguez-Bolaños et al. (2000).

43. Safina C (2020) Becoming wild: How animal cultures raise families, create beauty, and achieve peace. Henry Holt and Co., New York, NY

44. Gompper ME (2004) Correlations of coati (*Nasua narica*) social structure with parasitism by ticks and chiggers. *In*: Sánchez-Cordero V, Medellín RA (eds) Contribuciones mastologolicas en homenaje a Bernardo Villa. Instituto de Biología e Instituto de Ecología, UNAM, México, D.F., pp 527–534

45. Gompper ME (1994) The importance of ecology, behavior, and genetics in the maintenance of coati (*Nasua narica*) social structure. Ph.D. Dissertation, University of Tennessee, Knoxville

46. Based on observations in other studies (Gompper 2004; Hirsch et al. 2012); I observed little allogrooming.

47. Hirsch BT, Stanton MA, Maldonado JE (2012) Kinship shapes affiliative social networks but not aggression in ring-tailed coatis. PLoS ONE 7:e3730147.

48. Hass, CC (2003) Ecology of hooded and striped skunks in southeasthern Arizona. Final report submitted to the Arizona Game & Fish Department, Phoenix, AZ.

49. Clutton-Brock TH, Gaynor D, McIlrath GM, Maccol ADC, Kansky R, Chadwick P, Manser M, Skinner JD, Brotherton PNM (1999) Predation, group size and mortality in a cooperative mongoose, *Suricata suricata*. Journal of Animal Ecology 68:672–683

50. Wrangham RW, Rubenstein DI (1986) Social evolution in birds and mammals. *In*: Rubenstein DI, Wrangham RW (eds) Ecological aspects of social evolution. Birds and mammals. Princeton University Press, Princeton, N.J., pp 452–470

51. Clutton-Brock TH (2006) Cooperative breeding in mammals. *In*: Kappeler PDPM, Schaik PDCP van (eds) Cooperation in Primates and Humans. Springer, Berlin, GER, pp 173–190

52. Rood JP (1986) Ecology and social evolution of mongooses. *In*: Rubenstein DI (ed) Ecological aspects of social evolution. Birds and mammals. Princeton University Press, Princeton, NJ, pp 131–152

53. Komdeur J, Eikenaar C, Brouwer L, Richardson DS (2008) The evolution and ecology of cooperative breeding in vertebrates. *In*: Encyclopedia of life sciences. John Wiley & Sons, Ltd, Chichester, UK, pp 1–8

54. Moehlman PD (1986) Ecology and cooperation in canids. *In*: Rubenstein D, Wrangham R (eds) Ecological aspects of social evolution. Birds and mammals. Princeton University Press, Princeton, N.J, pp 64–86

55. Solomon NG, French JA (1997) The study of mammalian cooperative breeding. *In*: Solomon NG, French JA (eds) Cooperative breeding in mammals. Cambridge University Press, Cambridge, UK, pp 1–10

56. Gilchrist JS (2007) Cooperative behaviour in cooperative breeders: costs, benefits, and communal breeding. Behavioural Processes 76:100–105

57. Carl Bock, personal communication.

8

Costs of Sociality

Although group living in coatis provides benefits such as those described in the previous chapter, social living also has its costs. Diseases and parasites transmit much faster through a group. Loss of group members, particularly older females, means loss of information about local resources and may make groups more vulnerable to predation. Most studies have focused on the benefits of group living, to try to understand why coatis are more social than their relatives. Here I outline some of the costs of sociality, to try to understand why some coatis live in groups and others do not.

Parasitism and Disease

Coatis are vulnerable to canine distemper and rabies. In 1960, a distemper epidemic decimated the coati population in the Huachucas.[1] Although one of my radiocollared coatis died from distemper, I saw no evidence of it in any other animals. I attribute this to the fact that most dog owners, at least north of the U.S.-Mexico border, now vaccinate their dogs against canine distemper, so there may be less exposure to the disease. Southeastern Arizona is home to enzootic rabies, meaning that at any given time, some animals, usually gray foxes or skunks, are infected with the disease.[2] Reports of rabid coatis in Arizona surface once every few years, with ten reported cases between 1967 and 2012.[3,4] There was also a possible epidemic reported in Cancun in 2008,[5] but overall, rabies is uncommon for such a social animal.

Studying carnivorans in southern Arizona meant exposure to rabies (not necessarily in the clinical sense). To determine the possible exposure rate of coatis and other carnivorans to rabies, I sent blood samples from captured animals and brain samples from animals that had died to the Centers for Disease Control (CDC) in Atlanta, Georgia. From the brain samples, they could determine if the virus was present, and what variant it was. There were several variants (strains) of rabies in the area: the south-central skunk variant, the Arizona gray fox variant, and several bat variants. From the blood samples, they could determine if the animal had circulating antibodies to the rabies virus. However, because the rabies virus travels via the nerves, animals could become infected with the virus and die before they developed blood antibodies. With extremely rare exceptions, mammals are not carriers of rabies – a carrier can transmit the disease without succumbing to it themselves. More than 99% of the time, when an animal contracts rabies, it dies.[6] Of 82 blood samples and three brain samples of coatis, none had any rabies virus or antibodies. However, one gray fox had serum antibodies, and two skunks had virus in their brain tissue.[2,7]

I encountered rabies several times while in the field. In the first instance, I had captured a gray fox in a box trap set for coatis. Most trapped animals react when approached in a trap, usually by vigorously trying to escape. But this fox calmly lay in the trap, casually watching me prepare the handling equipment. I was able to drug him without a fight. Other than his oddly serene behavior, he appeared healthy. In addition to applying an ear tag, I took a blood sample and returned him to the trap to recover and released him a few hours later. His blood sample came back positive for rabies antibodies.

Several years later, I was driving up Garden Canyon, and passed a couple of Military Police officers driving down the canyon. I glanced in my side mirror, as I often do (it is amazing how many interesting things happen just behind you), and saw their brake lights go on. I slowed and watched a gray fox come around from the front of their jeep, trot alongside and suddenly bite at the rear tire of their vehicle. I hit the brakes to watch, as this seemed like rather aberrant behavior. The fox aggressively attacked the tire for about a minute, and then the MP's (who must have been watching their side mirror) slowly pulled the vehicle forward. The fox let go, and trotted up the bank next to the road, where it sat and looked at us. The MPs and I got out of our respective vehicles, exchanging looks of amazement and "did you see that?" questions. A moment later, they got in their vehicle and drove off. I watched them leave with dismay, as that left me alone with an obviously deranged fox, who was sitting quietly on the bank watching me. My curiosity got the better of me, and I slowly approached it to see if it was wearing one of my ear tags. It was. It also had a round hole through one ear that I recognized, so that even though I could not read the ear tag, I knew I would be able to identify it from my notes. It turned out to be an adult female that I had captured about seven months earlier. She also looked quite healthy, with bright shiny eyes and a luxurious coat.

Curiosity satisfied, I turned to return to my truck. Just then, she trotted down the bank, came over and grabbed the shoelace on my right boot in her mouth. She started tugging and growling like a puppy. I froze, heart pounding, as I tried to figure out what to do. This animal was obviously not right in the head, which is one of the symptoms of rabies. On that day, I was not doing any trapping and did not have any knock-out drugs in my truck. I carried no weapons, and the only thing I could think to do was stomp on her head. But for the moment, she was busy with my shoelace. I worried that any quick movement on my part might cause her to become viciously aggressive, perhaps going after my face or hands. So, I slowly dragged her back to my truck, quietly opened the door and slid inside, then carefully bounced the truck door on her head. She finally detached herself from my foot and trotted off into the woods. I sat trembling in my truck for several minutes before I continued up the canyon to look for coatis. Even though I was vaccinated against rabies, it is always scary to confront an animal (or human) that is acting unpredictably, and I could have been seriously injured if she had decided to bite me. She was never seen again, so it could not be confirmed that she had rabies, but when I sent a query to the Arizona State Health Department, they responded that the behavior I observed was not unusual for rabid gray foxes.

Another encounter with rabies occurred when I was radiotracking skunks, as part of my study of the ecology of striped and hooded skunks in southeastern Arizona. The radio collars I used on the skunks had an activity sensor which broadcast a chattering sound when the animal was moving, only audible to my telemetry receiver. In this case, I had been receiving an active signal from one female hooded skunk as I was getting the locations of the other radiocollared skunks that were sound asleep in their dens. This particular female was often active until well after daylight, but on this day, she was still active as it was approaching noon. I set off to see what she was up to. As I hiked up a hillside on the west side of her home range, I approached a large boulder where her signal seemed to be coming from. As I got near, I could hear the stomp and hiss of a skunk. Assuming the noise came from her, I set down my radio antenna and took out my note pad to record the location. Suddenly the female hooded skunk ran around from the far side of the boulder straight toward me. I grabbed my gear and bolted down the hill away from her. A dozen or so meters down the hill, I stopped to look back. She had changed direction and was moving along the hill to the north. She moved rapidly, running to the base of the oak trees, quickly digging small holes as if looking for insects, then moving on to the next tree. While keeping an eye on her, I moved back up to the large boulder to see what had hissed and stomped at me. When I peered under the boulder, I saw a rather agitated spotted skunk. I then turned and followed the hooded skunk down the hill, as she continued in her strange foraging pattern. I was not sure what to make of it. Rabid animals often lose their appetites, so the fact that she was foraging was not consistent with rabies. However, her daytime activity and very rapid movements were. I had trouble keeping up with her, as she ran from tree to tree. I followed her for about 30 minutes, and then went on to try to check on other radiocollared skunks.

When I returned to the field a couple of days later, a mortality signal led me to her carcass more than two miles north of her usual home range. It appeared a raptor or owl killed her. I collected her skull, and later sent some brain tissue to the CDC. It came back positive for rabies.[2] I also found several dead hooded and hog-nosed skunks over the years; I had one tested for rabies and it came back positive. Most of the dead skunks had a characteristic posture – they were splayed out, with their hind limbs stretched behind them, and fore limbs stretched in front. Rabies can result in paralysis, starting in the hind end, so it is possible these animals were dragging themselves along with their front legs before they died.

Jack Childs, of the Borderlands Jaguar Detection Project, reported to me an encounter with a coati that might have had rabies. On June 9, 2004, Jack and his wife were out looking for jaguar tracks near the U.S.-Mexico border, when they ran across a male coati near a cattle watering trough. The coati appeared disoriented, and was making a loud moaning sound, almost like a scream. As they watched him, the coati turned and started coming toward them. They tried to move out of the way, but he kept coming and started getting extremely aggressive. Unable to outrun the coati, or chase him away, Jack was finally forced to bludgeon him to death with his walking stick. Unfortunately, they were unable to get the coatis' brain tested for presence of rabies virus, but the behavior is consistent.

Rabies occurs in species-specific variants that maintain reservoirs within particular hosts. It will occasionally spill over to other species but does not transmit as well as it does among their own hosts. It is most common in bats and carnivorans, but any mammal can contract it.[8] It is typically spread when an animal infected with rabies has the virus in its saliva and bites another mammal. Within coati habitat near the U.S.-Mexico border, the primary rabies reservoirs are bats, skunks, and foxes. The raccoon rabies that is present in the eastern U.S. and southern Canada is not currently found in areas where coatis live. Raccoon rabies is widespread in the eastern U.S., and the U.S. Government (through USDA-APHIS Wildlife Services) is attempting to contain the spread by using oral baits containing rabies vaccine.[8] Due to the low and patchy density of raccoons in coati habitat within the U.S.,[9] it is unlikely that raccoon rabies would make much headway there, but a potential avenue might be through southern Texas to Tamaulipas. No reports of coatis contracting raccoon rabies are available, so it is not known how vulnerable they would be to that variant.

In Brazil, and some other tropical environments, coatis are significant hosts for *Trypanosoma cruzi* – the parasite that causes Chaga's disease, a debilitating disease that infects humans and some livestock. Humans are infected when the triatomine bugs (including kissing bugs) bite and then defecate on their victim. Transmission occurs when the parasite-infested feces are accidentally rubbed into mucous membranes or breaks in the skin. Coatis, raccoons, and other carnivores host a variety of strains of the parasites and may serve as reservoirs that provide the triatomine bugs with the parasite.[10] Some studies have found that female coatis have a much higher rate of parasitism than do males, but rather than being transmitted socially, it appears that the bugs may be infesting arboreal natal nests, where female coatis stay with their young for 6-7 weeks.[10-12] The parasite appears to have little effect on the coatis' health, although infection by a related parasite, *Trypanosoma evansi*, which is transmitted by the bites of some flies, appears to cause anemia.[12] Chaga's disease occurs throughout the American tropics, through Mexico and into the southern U.S., where its hosts appear to be raccoons and opossums. No other studies have examined the prevalence of *T. cruzi* in coati populations outside of South America.

In Mexico, Valenzuela euthanized 39 coatis to reduce a scabies epidemic after capturing seven animals dying of the disease,[13,14] indicating that in some populations, contagious parasite-induced diseases can be a cost of sociality. He also found that ectoparasites, including the mites that cause scabies, were more prevalent in a population that had a higher density of coatis. Few outbreaks of disease or parasitism have been documented in other studies, and it appears that these costs of sociality are outweighed by the benefits, and by the coatis' high reproductive rate.

Group Dynamics and the Allee Effect

As mentioned previously, predation rates were higher on groups with fewer than four adult females. Very few troops with three or fewer adult females persisted for long;[15] in some cases, the smaller troop joined an existing larger troop, and in other cases, I lost track of them entirely. I do not know if they traveled a larger distance to join another troop or died out completely.

The population of coatis in the Huachucas declined rapidly during my study. By 2000, I only had two troops remaining with radio collars, although there were still reports of coatis in Ramsey Canyon which may have been what was left of Troop 1. One radiocollared female remained with the troop, but the battery on the collar was dead. As the total population declined, small troops of coatis scrambled to join other troops, so that the number of troops declined faster than overall troop size. The average troop size stayed at or above 15 animals until 1999, when only four of my original nine troops remained. But finally, toward the end, it may have been simply too difficult to find another troop to join. Based on sighting reports I received from hikers and birders visiting other areas in the Huachucas, this decline appeared to include the entire mountain range, not just my study area. McColgin reported a similar decline in the Chiricahua Mountains,[16] 70 miles (113 km) away from the Huachucas.

One of the saddest periods for me during the entire study was the summer of 2000. Just before Troop 3 split up for nesting, there were five radiocollared females in the group, although only three of the radio collars were still functioning. Right about the time they were to have given birth,

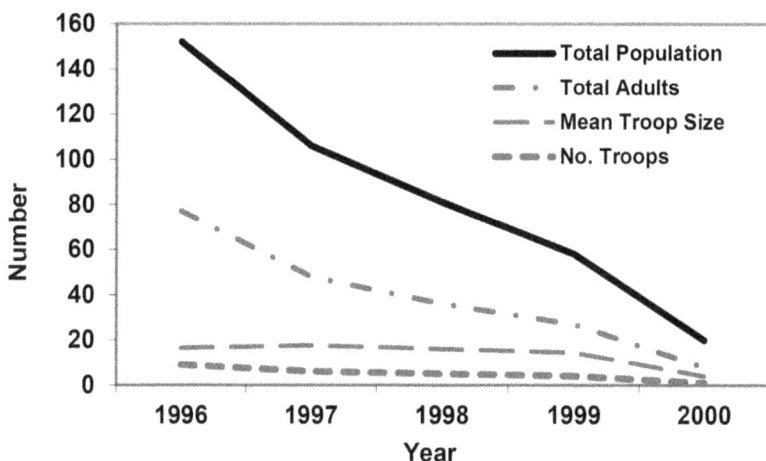

The decline in coati numbers in the Huachuca Mountains study area, 1996–2000. The graph reflects only the animals I monitored, not the entire population of the mountain range. However, the decline was range-wide, based on reported sightings (and lack thereof).

F30 and F167 were killed by mountain lion(s) within a couple of days of each other. Both were still pregnant when I last saw them, but I do not know if they had given birth yet before they were killed. Suddenly, I was down to one female with an active radio collar, F69, who I had been tracking since 1996. With no other coatis to track, I focused all my attention on her, trying to keep her in sight as she climbed up agave stalks to feed on the nectar, dug through the leaf litter, and pounced on crickets. As the females usually did, right before she brought her litter out of the nest, she started moving much longer distances, probably looking for her troop mates. She brought her five kits out on the 5th or 6th of August, and I observed them on the west side of her home range on August 7th.

For the next several weeks, I located her throughout her home range, always alone with her kits. I had hoped that one of the females wearing a non-functional radio collar might still be alive, and that they would find each other. I found her for the last time on August 22, at the very northern end of the troop's home range. Her signal disappeared after that, but I never found out if it was due to battery failure (which was overdue), or she left the area. The silence from the telemetry receiver was heartbreaking as I hiked hundreds of miles and searched in wider and wider circles over the entire mountain range, but never heard her signal or saw her again. As I searched for her, I could not help but think back to many encounters

F69 foraging in a Parry's agave, July 2000.

I had had with F69. I had captured and radiocollared her in the summer of 1996 and watched as she brought her new litter back to the troop each year. Her unique, high-pitched chirp let me know if she was in the group, where I usually found her surrounded by at least 10 youngsters, appearing to take on the role of babysitter.

I ran across hardly any coati sign during my searches, and no indications of any troops remaining on Fort Huachuca. I had lost contact with Troop 5 in May when F185, the last active radio collar, was killed by a mountain lion. During July, I had observed F149, also from Troop 5 on several occasions, in an area where she had nested before in lower Garden Canyon. Unfortunately, her collar was no longer active, and I never found out what became of her or her kits.

That summer was sad for another reason. Mike Seidman was with me on that day when I received the last signal from F69. Mike had volunteered on my coati project since the beginning, traveling from Phoenix about once a month to spend the night in the Huachucas and help keep track of the coatis. Mike was an avid hiker, and I usually had trouble keeping up with him. However, on that day, he seemed a bit tired, and we often paused to rest as we climbed the steep trails in F69's home range. A few months later, he was diagnosed with inoperable liver cancer and he died at the end of 2002. Mike's love of the coatis and curiosity about them was a continual reminder that what I was doing was worthwhile, and I was honored to be on one of his last hikes in the Huachucas. He was also an ardent voice for conservation in Arizona, a voice that is badly missed.

Sightings of coatis throughout the range had plummeted. As part of a subsequent skunk study on Fort Huachuca from 2001-2003, I operated trail cameras in the canyons on the Fort, only photographing two male coatis (or one male twice) during the entire time. I started giving presentations about the coati study, imploring people to provide any information about coatis they had seen in the Huachucas. Sightings were few and sporadic, with only a few coatis seen on or near the Fort until 2007, when small groups were seen at several locations in the mountain range. These sightings represented a huge decrease from the mid-1990's, when large groups of coatis were reportedly almost weekly. The population of coatis in the Huachucas was almost gone. As of this writing (summer 2020), coati sightings are finally becoming more common in the Huachucas and may finally be rebounding to their earlier levels. So, it took about 15 years for the population to recover, which is quite remarkable for an animal with such a high reproductive rate (each female produces about four offspring every year).

The rapid decline of the coatis illustrates how vulnerable they are when their populations get low. Although they may forage more efficiently alone (perhaps depending upon what they are foraging for), they are much more vulnerable to predators. So, in areas with large predators that have keyed into coatis, troops of fewer than 10 animals are extremely vulnerable to predation and may ultimately disappear if they cannot find another troop to join.

This illustrates an often-overlooked principle in ecology. In general, we assume that population growth rates are relatively constant until resources, particularly food, become limiting. As numbers of coatis decline, we might expect mountain lions to have more difficulty finding them and start seeking alternative prey. But coatis, at least at the northern edge of their range, are not the mountain lion's primary prey, they are secondary prey. Mountain lion numbers are maintained by the numbers of white-tailed deer, mule deer, and javelina. If the area is inhabited by one or more mountain lions that have acquired a knack for hunting coatis, we can expect predation rates on coatis to remain at least constant, if not increase, as the coati population declines. Meanwhile, once the coati troops number four or fewer adult females, they become more vulnerable, and predation rates increase. This increased mortality rate at low numbers is known as the "Allee effect."[17-19] Recent analysis has shown that populations of a rare species that are the secondary prey of a common predator are extremely vulnerable to catastrophic population crashes,[20-22] which is what I think occurred in the Huachucas beginning in 2000. Social species that require a minimum number of animals to survive are particularly vulnerable, and thus sociality itself may also become a cost.[21,23,24]

The population in the Huachucas suffered from reduced juvenile survival, with an average of fewer than one juvenile per female surviving winters in the late 1990's. The lower the winter precipitation, the lower the overwinter survival of juveniles, and four of the winters between 1995 and 2000 had less than half of the long-term average of winter precipitation. Because abundance of invertebrate prey is correlated with soil moisture (see Chapter 13), dry winters may reduce food availability. Mortality rates continued to be high on subadults, resulting in 85-97% mortality by the time the coatis were 17 months old.[25,26] But the population was also subject to high predation rates on the adults, and some predation on the juveniles. Reduced survival of juveniles, subadults and adults was more than the population in the Huachucas could bear. This high level of predation has not been reported in other studies, but most studies have been in areas with lower predator densities.[27-29]

But the story is made more interesting by the fact that the population collapse appeared restricted to the Huachucas. I received sighting reports from most of the other mountain ranges in southeastern Arizona and a few in New Mexico. Troop sizes in these other areas were much reduced, with most sightings of 8-15 animals, down from sightings of 20-30 coatis during the mid-1990s.[16] I think the widespread decline was due to several dry winters in a row that reduced the overwinter survival of juveniles, much as occurred in my study area. Populations in these other areas, however, did not crash the way they did in the Huachucas.

When I have presented the above observations on coati mortality related to mountain lion predation to various audiences, I can almost feel audience members bristle at my mention that much of the population decline was due to predation. I think some of this stems from a lack of

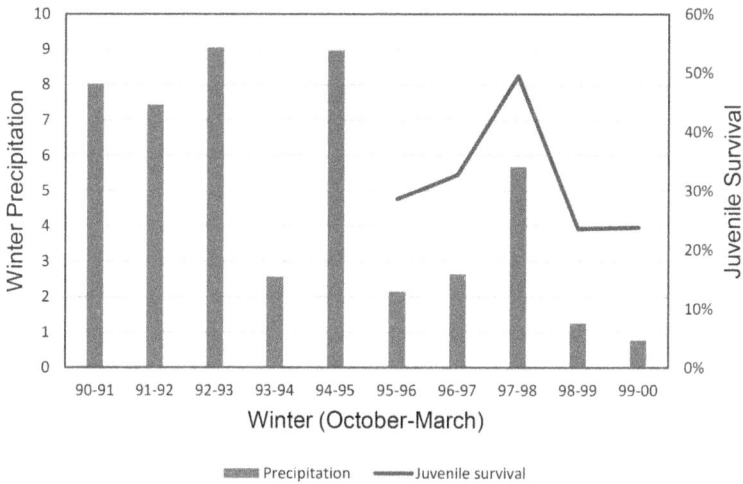

Winter (October to March) precipitation and overwinter survival of juvenile coatis (September to April). Percent juvenile survival calculated from the difference between the potential juvenile:adult female ratio to the observed ratio. Precipitation recorded at the Central Meteorological Office, near Libby Army Airfield, Fort Huachuca

understanding of populations dynamics (yes, in some cases predators can control their prey), and that people may be assuming that I am pushing for control (harvest) of mountain lion populations. I consider mountain lion predation on coatis a completely natural event, and I would not advocate for killing mountain lions to protect coati populations. Coatis evolved with big cats. The coatis in the Huachucas probably could have tolerated high levels of predation if winter rains had been more abundant.

Coatis in Central America are also vulnerable to periodic droughts. The same El Niño events that bring increased rain to the American southwest result in droughts in Central America. Population crashes of coatis in Panama have been attributed to decreased fruit production from these droughts.[29,30] During Kaufmann's study, he also found that older females were not producing litters, and some females waited until three or four years of age to produce their first litter; both factors may be related to reduced food availability.[31]

Cultural Transmission – the Downside

The local extinction of troops takes with it the accumulated knowledge of the elders – where to find water (critical during the dry season), where to find trees when they are producing fruit, where the good hiding places are. As Carl Safina writes,

> *Culture comprises knowledge and skills that travel from individual to individual and generation to generation. It is learned socially. Individuals pick it up from other individuals. It is knowledge that doesn't come from instinct alone.*[32]

Coatis do not appear to have strong dispersal tendencies, so it may take quite a while for habitats to be recolonized. If large predators are still in the area, it may take even longer, as we can assume colonizing groups will probably be small enough to be vulnerable to predation. Establishment of new troops at the northern end of their range may take a fortuitous combination of events, including a large enough starting population (or no predators) and ample winter rains for the first few years. New cultures must be established. Conservationists are starting to recognize the importance of unique cultures when saving populations of animals.[33]

Spending all day alone, hiking around trying to pick up signals of radiocollared coatis, gives a person a lot of time to think. I spent some of that time pondering why coatis had this strange social system. Given the high rate of mortality I was seeing on the radiocollared males, it seemed like it would benefit males to remain in their natal troops or join a neighboring troop. So why did they leave the troop to take up a solitary existence? There have not been any observations suggesting that adult females in the troops would drive out the males as they reached adulthood, rather it appears that they left of their own volition. I wondered, initially, if they did not just become surly as they got older, eschewing the company of males and females alike. The agonistic behavior between adult females and males outside of the mating season in both Arizona and Panama supported that idea.[16,31] But as results from other studies came in, and I had the opportunity to observe some captive coatis, I realized that it was not that simple. In some areas, adult males appeared to be well-integrated into the troops - grooming and being groomed by adult females, playing with juveniles and subadults. Captive adult males can be quite social and friendly with each other. So, they appear to retain their social "manners." This makes them even more unique: not just a rare social system with group-living females and solitary males, but a system in which the sociality of males is flexible. Unfortunately, few populations of coatis have been studied long enough to examine the role of cultural learning in these social arrangements. Is it just a matter of resources, or could the propensity for some males to be more social be a cultural phenomenon?

In analyzing the costs and benefits of sociality in coatis, we must keep in mind that there are both social (female) and less social (male) coatis. When and how often a male associates with a troop may be an analysis of costs and benefits. He might benefit from reduced predation risk. He might benefit from occasional grooming to reduce parasites. On the other hand, he might have more competition for food, increased exposure to parasites, and constant badgering by adult females and juveniles. This social/asocial arrangement is unique among carnivorans, and only approximated by elephants and sperm whales.[34,35] However it appears that male coatis may be even more flexible in their sociality than these other species. This arrangement differs markedly from the coatis' cousin, the raccoon.

Among raccoons, the most long-lasting social bonds appear between a mother and her offspring. The offspring typically disperse by their first birthday, at which time both males and females are capable of breeding – at least a year earlier than coatis. Female offspring disperse only a short distance and may overlap home ranges with their mothers.[36,37] Males tend to disperse much farther, up to 165 miles (265 km).[38] Ironically, given their ubiquity near human habitation, social behavior in raccoons is less studied than in coatis. Only recently have detailed studies of raccoons been conducted, and they have found that adult male raccoons sometimes form social groups of three or four individuals that last for considerable periods of time, even though they may not be closely related.[39-41] Of course, raccoons are known to congregate at rich food sources, such as trash dumps, but these appear to be just congregations for feeding, with no social bonds among animals.[38] Related animals may den together during the winter, indicating that thermoregulation plays a role in raccoon social behavior.[42] More research on sociality in raccoons and other procyonids may provide insight into how this interesting group of animals uses multiple solutions to surviving in similar habitats.

Notes and References

1. Risser SC Jr (1963) A study of the coati mundi *Nasua narica* in southern Arizona. Master's Thesis, University of Arizona, Tucson
2. Hass CC, Dragoo JW (2006) Rabies in hooded and striped skunks in Arizona. Journal of Wildlife Diseases 42:825–829

3. J.W. Krebs, CDC, (personal communication), Jack Childs (personal communication). Arizona Dept. of Health Services reports, 1999-2010.

4. Krebs JW, Williams SM, Smith JS, Rupprecht CE, Childs JE (2003) Rabies among infrequently reported mammalian carnivores in the United States, 1960-2000. Journal of Wildlife Diseases 39:253-261

5. Arechiga-Ceballos N, Velasco-Villa A, Shi M, Chavez S, Barron B, Cuevas-Dominquez E, Gonzalez-Origel A, Aquilar-Setien A (2010) New rabies virus variant found during an epizootic in white-nosed coatis from the Yucatan Peninsula. Epidemiology and Infection 138:1586-1589

6. Hanlon CA, Childs JE, Nettles VF, the National Working Group on Rabies Prevention and Control (1999) Rabies in wildlife. Journal of the American Veterinary Medical Association 215:1612-1618

7. Hass CC (2003) Ecology of hooded and striped skunks in southeastern Arizona. Final report submitted to the Arizona Game & Fish Department, Phoenix, AZ. Phoenix, AZ

8. Birhane MG, Cleaton JM, Monroe BP, Wadhwa A, Orciari LA, Yager P, Blanton J, Velasco-Villa A, Petersen BW, Wallace RM (2017) Rabies surveillance in the United States during 2015. Journal of the American Veterinary Medical Association 250:1117-1130

9. Personal observation.

10. Rocha-Mendes FL, Roque ALR, de Lima JS, et al (2013) *Trypanosoma cruzi* infection in Neotropical wild carnivores (Mammalia: Carnivora): at the top of the *T. cruzi* transmission chain. PLoS ONE 8:e67463

11. Herrera HM, Lisboa CV, Pinho AP, Olifiers N, Bianchi RC, Rocha FL, Mourão GM, Jansen AM (2008) The coati (*Nasua nasua*, Carnivora, Procyonidae) as a reservoir host for the main lineages of *Trypanosoma cruzi* in the Pantanal region, Brazil. Transactions of The Royal Society of Tropical Medicine and Hygiene 102:1133-1139

12. Olifiers N, Jansen AM, Herrera HM, Bianchi R de C, D'Andrea PS, Mourão G de M, Gompper ME (2015) Co-infection and wild animal health: effects of Trypanosomatids and gastrointestinal parasites on coatis of the Brazilian Pantanal. PLoS One 10:e0143997

13. Valenzuela D (1998) Natural history of the white-nosed coati, *Nasua narica*, in the tropical dry forests of western Mexico. Revista Mexicana de Mastozoologica 3:26-44

14. Valenzuela D, Caballos G, Garcia A (2000) Mange epizootic in white-nosed coatis in western Mexico. Journal of Wildlife Diseases 36:56-63

15. Hass CC (2002) Home-range dynamics of white-nosed coatis in southeastern Arizona. Journal of Mammalogy 83:934-946

16. McColgin ME, Koprowski JL, Waser PM (2018) White-nosed coatis in Arizona: tropical carnivores in a temperate environment. Journal of Mammalogy 99:64-74

17. Courchamp F, Clutton-Brock TH, Grenfell B (1999) Inverse density dependence and the Allee effect. Trends in Ecology and Evolution 14:405-410

18. Odum EP (1959) Fundamentals of ecology. W.B. Saunders Company, Philadelphia, PA

19. Amarasekare P (1998) Allee effects in metapopulation dynamics. The American Naturalist 152:298–302

20. Courchamp F, Grenfell B, Clutton-Brock T (1999) Population dynamics of obligate cooperators. Proceedings: Biological Sciences 266:557–563

21. McLellan BN, Serrouya R, Wittmer HU, Boutin S (2010) Predator-mediated Allee effects in multi-prey systems. Ecology 91:286–292

22. Sweitzer RA, Jenkins SH, Berger J (1997) Near-extinction of porcupines by mountain lions and consequences of ecosystem change in the Great Basin Desert. Conservation Biology 11:1407–1417

23. Clutton-Brock TH, Gaynor D, McIlrath GM, Maccol ADC, Kansky R, Chadwick P, Manser M, Skinner JD, Brotherton PNM (1999) Predation, group size and mortality in a cooperative mongoose, *Suricata suricata*. Journal of Animal Ecology 68:672–683

24. Sherman PW, Runge MC (2002) Demography of a population collapse: the northern Idaho ground squirrel (*Spermophilus brunneus brunneus*). Ecology 83:2816–2831

25. Hass CC, Valenzuela D (2002) Anti-predator benefits of group living in white-nosed coatis (*Nasua narica*). Behavioral Ecology and Sociobiology 51:570–578

26. Note that I cannot separate dispersal from mortality here, but the result for the natal troop is the same.

27. Binczik GA (2006) Reproductive biology of a tropical procyonid, the white-nosed coati. Ph.D. Dissertation, University of Florida, Gainesville

28. Hirsch BT (2007) Within-group spatial position in ring-tailed coatis (*Nasua nasua*): balancing predation, feeding success, and social competition. Ph.D. Dissertation, Stony Brook University, Stony Brook, NY

29. Gompper ME (1997) Population ecology of the white-nosed coati (*Nasua narica*) on Barro Colorado Island, Panama. Journal of Zoology, London 241:441–455

30. Wright SJ, Carrasco C, Calderon O, Paton S (1999) The El Niño southern oscillation, variable fruit production, and famine in a tropical forest. Ecology 80:1632–1647

31. Kaufmann JH (1962) Ecology and social behavior of the coati, *Nasua narica*, on Barro Colorado Island, Panama. University of California Publications in Zoology 60:95–222

32. Safina C (2020) Becoming wild: How animal cultures raise families, create beauty, and achieve peace. Henry Holt and Co., New York, NY

33. Brakes P, Dall SRX, Aplin LM, et al (2019) Animal cultures matter for conservation. Science 363:1032–1034

34. Booth-Binczik SD (2001) Ecology of coati social behavior in Tikal National Park, Guatemala. Ph.D. Dissertation, University of Florida, Gainesville

35. MacDonald DW (ed) (1984) The encyclopedia of mammals. Facts on File, New York, NY

36. Gehrt SD, Fritzell EK (1998) Duration of familial bonds and dispersal patterns for raccoons in south Texas. Journal of Mammalogy 79:859–872

37. Ratnayeke S, Tuskan GA, Pelton MR (2002) Genetic relatedness and female spatial organization in a solitary carnivore, the raccoon, *Procyon lotor*. Molecular Ecology 11:1115–1124

38. Zeveloff SI (2002) Raccoons: a natural history. Smithsonian Books, Washington, D.C.

39. Chamberlain MJ, Leopold BD (2002) Spatio-temporal relationships among adult raccoons (*Procyon lotor*) in Central Mississippi. The American Midland Naturalist 148:297–308

40. Gehrt SD, Gergits WF, Fritzell EK (2008) Behavioral and genetic aspects of male social groups in raccoons. Journal of Mammalogy 89:1473–1480

41. Hauver SA, Gehrt SD, Prange S, Dubach J (2010) Behavioral and genetic aspects of the raccoon mating system. Journal of Mammalogy 91:749–757

42. Robert K, Garant D, Vander Wal E, Pelletier F (2013) Context-dependent social behaviour: testing the interplay between season and kinship with raccoons: Social interactions, seasonality and kinship. Journal of Zoology 290:199–207

9

Perception and Communication

I n October 1994, I went down to the Huachuca Mountains on a recon-
naissance mission prior to the start of my study. One day on that trip,
I took a hike up Ramsey Canyon to try to find a troop of coatis that
were often seen there. I hiked through the Nature Conservancy Preserve
and up to a rocky promontory, which gave an incredible view up and
down the canyon. The canyon bottom was lit up with color – scarlet red
maples, and golden sycamores and walnuts, in stunning contrast to the
deep greens of the oak, pine, and juniper-covered hillsides. I turned and
started back down the switchbacks, meeting a couple of people hiking on
their way up. They reported having just seen a group of animals that
they described as crosses between anteaters and monkeys. I quickly ex-
plained to them what they saw and obtained more specific details on the
location of their sighting before rushing down the trail.

When I arrived at the spot they described, I pulled a small camcorder
out of my backpack, crawled up on a rock, and waited. Within a few minutes, I could hear many small feet crunching through the leaf litter, accompanied by a variety of grunts, chirps, and squeaks. I videotaped the
coati troop as they approached, foraging and squabbling over found
morsels. Youngsters chased each other up and down trees, and when she
finally discovered me, the one of the elders climbed a nearby tree and
barked at me for several minutes. Finally, the troop ran single file across
the trail and disappeared into the forest. The 16 minutes of video I

obtained was the longest, most complete collection of behaviors I would ever be able to tape during the entire study. I just happened to be at the right place at the right time.

Collecting Information

Coatis are creatures of the forest and scrub – thick, brushy environments with limited visibility, but full of rich smells. This is the world that they move through daily and the universe to which their sensory systems are adapted. To survive, a coati needs information about its environment: what to eat, where to find food and water, when and how to avoid predators, which troop mates are safe to play with and which ones to avoid. It gathers this information through a variety of senses: the sense of smell to find food, find and identify other coatis, and detect when predators are in the area. It uses the sense of sight to find food, detect danger, observe signals from other coatis, move through their environment and avoid dangerous objects. Coatis listen to detect predators and prey, as well as communicate with each other. Coatis use their sense of touch to find food, nuzzle and groom each other; and they use their sense of taste to determine which foods are palatable and which ones to avoid.

Compared to many other animals, human senses are quite limited. Although we see color quite well, we still see only a limited part of the electromagnetic spectrum. Other animals, such as owls, see in the infrared (lower wavelengths than we can see), and bees, moths, and some birds can see ultraviolet (higher wavelengths than we can see).[1] Some animals, including dogs, can see things moving at a higher flicker frequency than we can, which allows them to detect much faster motion, and is one reason they are so good at catching balls and disks in mid-air. Although we hear quite well, our hearing covers a relatively narrow portion of the sound spectrum, from about 20 Hz to about 20,000 Hz (or 20 kHz; for a teenager, most adults cannot hear above 15 kHz). Animals such as elephants and hippos can hear and communicate in infrasound, defined as sounds below the level of human hearing, and most mammals can hear sounds 2-5 times higher in frequency than we can (ultrasound). Many mammals can isolate sounds better, by being able to distinguish among tiny differences in frequency and call rate. And our sense of smell is quite pathetic, with animals such as bloodhounds having a sense of smell close to a million times better than ours. Our sense of touch is better in some respects, with the incredible sensitivity of our fingertips, but worse in others, as we have no

Coati primary sensory inputs: ears, eyes, nose, whiskers, paws. White-nosed coati, Arizona-Sonora Desert Museum.

vibrissae (whiskers) with which to feel our way through our environment. Different portions of our brains are dedicated to each sensory system. Humans have a large portion of their brains devoted to vision, whereas dogs have more of their brains devoted to smell, raccoons have a large portion of their brain devoted to touch sensors on their forepaws, and coatis have a large portion of their brains devoted to their nose.[2]

When I was out radio-tracking coatis, I would occasionally try to imagine how a coati perceives its environment (a human's sense of imagination being quite extraordinary). First, I would have to get low to the ground, as coatis are only about 30 cm (1 ft) tall. From this elevation, the perspective is much different. The smells on the ground are much closer and stronger. Although no one has measured a coati's sense of smell, based on the number of olfactory cells and the proportion of the brain devoted to olfaction, if I were a coati my sense of smell would be at least as good as a dog's and possibly better. In addition, my rhinarium, the tip of my nose, is highly flexible – I can bend it up and down and side to side. This allows me to orient my nose more precisely to localize smells, and I can stick my nose in small holes and cracks to sniff out my favorite prey – bugs! My muzzle is covered with long vibrissae, giving me a fine sense of touch along my

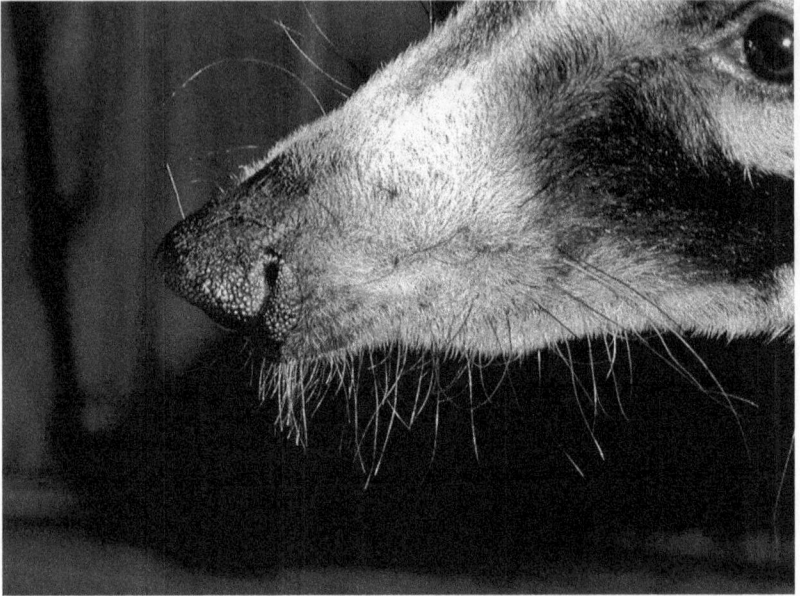

Dense vibrassae on the muzzle of a captive white-nosed coati.

muzzle, so I can feel bugs moving in the soil and leaf litter. I can tell my troop mates apart by their individual odors.

The closeness of the ground and the thickness of the vegetation also alter the acoustics, as sounds reflect off these surfaces. This not only effects what I can hear, but also shapes my vocal repertoire.[3] My small ears mean that I probably cannot detect quiet or distant sounds like a deer or jackrabbit can with their oversized ears. However, as a coati, I have an incredibly broad range of hearing – I can hear up to 95 kHz – much higher than humans 20 kHz,[4,5] and even much higher than a dog's 60 kHz. My total bandwidth of sensitivity covers 7½ octaves, among the highest for carnivorans, with my most sensitive hearing occurring between 0.25 and 45 kHz. What am I hearing in those high frequencies? Well, researchers do not know for sure, but many insects and rodents communicate that high. The sound of rustling leaves also extends into high frequencies, so, as a coati, I might be listening for prey or the approach of predators. It's also much easier to determine the direction of high frequency sounds.[6] And, as will be discussed later, I emit sounds in high frequencies, so I may be just listening to other coatis. Higher frequency sounds attenuate (lessen) much more with distance and in vegetation, so these sounds do not travel very far in forested habitats.[7] High frequency communication, therefore, gives my troop mates and me a way to whisper to each other with less worry about predators eavesdropping. My feline predators can also hear up in these high frequencies (up to 65 kHz),[8] but due to the nature of sound in the forest, only if they are very close.

Visibility is limited to a few feet or few yards in the forest and jungle. The bushes and trees seem much taller than to a human, but my small body size makes squeezing through the brush much easier. If I were a coati, the forest would look quite different. I could see yellow and red and blue just fine, but I could not distinguish green from gray or brown very well.[9] This means I have no trouble seeing ripe fruits and berries, or even brightly colored insects and snakes, but the myriad shades of green that people see when they look at the forest are shades of gray to me. The rusty-brown color of my troop mates might stand out against the gray of the forest. My eyes face forward – not as much as a primates' eyes do, but enough to give me stereoscopic vision for navigating through the upper branches of the trees. Although some researchers consider coati vision to be rather poor, I have spotted humans from at least 50 meters.[10]

Communication

Coatis communicate with each other using a variety of visual displays, vocalizations, touches, and odors. Each of these methods of communication has distinct advantages and disadvantages relating to the type of information that can be conveyed and limitations due to time, distance, the perceptual abilities of the receiver, and the influence of the habitat on the signal itself. I concur with other biologists in defining communication as an intentional transfer of information that involves a signaler, a signal, and a receiver.[1,11-14] Some biologists suggest that communication involves manipulation – that is, the purpose of sending a signal is to evoke a response from the recipient, rather than simply transfer information.[1,15] The response from the recipient may not be dramatic, and may not even be visible to us, but if it changes the emotion or future actions by the recipient, then the signaler is considered to have manipulated the receiver.

Information transfer also occurs in groups, where the sender may transmit a signal to the entire group (at least those that are paying attention), such as when an animal issues an alarm call. Such signals are easily eavesdropped on, that is, animals that are not the intended recipients are able to collect information from such signals. For example, California ground squirrels increased vigilance after hearing alarm calls of California quail or song sparrows, but not after hearing their songs.[16] Animals may also get information about the quality of food sites by watching other animals feeding or foraging; these are examples of "public information."[12,17] Note that neither of these last two types of information gathering technically qualify as communication, because they lack the intentional sender--signal--receiver chain.

In addition to watching wild coatis, I obtained close-up observations and recorded the vocalizations of two groups of captive coatis. Stan and Linda Rolinski rescued coatis for decades – taking in pet coatis that people wanted to get rid of for various reasons. Their coatis, up to a half-dozen or so, had free run of the house and yard, although an electric wire on the outer fence kept them from cruising the neighborhood. Most of the coatis were quite friendly and would readily climb in your lap to be cuddled and petted. The coatis interacted with each other often, and some of the males not only tolerated each other, but appeared to be good buddies, often playing and grooming each other – something you almost never see in the wild. I also took advantage of a similar group of coatis at the Arizona-Sonora Desert Museum in Tucson. These animals, much like the Rolinski's coatis, were former pets, and the group consisted of six adult males, including five white-nosed and one brown-nosed, and two adult females, both white-nosed. Although it was not a natural grouping, they did allow me to approach close enough for recording sounds and behaviors, and the staff generously allowed me to perform some playback experiments.

Animals often communicate using displays, which are intention movements, sounds, or odors that allow an animal to indicate that it is about to do something without actually doing it. Displays can indicate body size, territory ownership, and willingness to mate, play, or attack. By using displays instead of physical interaction, animals can reduce the amount of energy used as well as their risk of injury.[15]

Visual Displays

Visual displays are common methods of communication by birds and mammals. They have the advantage of being able to convey information over long distances, if the vegetation is not too thick. A lot of information can be included with a visual signal, including the identity and emotion of the sender. However, there are disadvantages, too. Visual signals do not linger – once the signal is done, it is gone. If you were not looking, or could not see the signal clearly, you probably did not get the message. A widely recognized visual signal is the wagging tail of a dog.

Coatis live in dense forests and brushy habitats where visibility is limited, and lighting is often dim. As mentioned previously, they can distinguish colors except for green and gray. They also have a tapetum in the back of their eyes, which reflects light onto the retina allowing them to see better in low light. Coatis do not appear to use many visual displays, probably because of the poor visibility in the habitats they occupy.

Facial and body coloring may serve as visual displays, by selectively drawing the eyes toward or away from specific body parts. Although many animals use color patterns to hide, for example, stripes to hide in the grass (tigers), or visually confuse a predator or parasite (zebras), or even startle a predator that has approached too close (eye spots on moths), others use markings to enhance existing ornamentation (facial stripes on some horned antelopes).[15] Coatis are quite drab in coloration except for the remarkable pattern of white spots on the face. White-nosed coatis have a white muzzle, whereas in the South American coatis the muzzle is brown or black, so muzzle color does not appear to be important by itself. However, both white-nosed and some brown-nosed coatis have the same pattern of white spots above, below their eyes, which may function to draw attention to the eyes or let another coati or predator know when the coati is facing it. Perhaps they make the coatis' eyes look larger, giving the appearance of a larger animal. The light tips of the ears would also increase their visibility, and visible ears are more likely to indicate a friendly coati, whereas ears that are back and out of sight show aggression. Little has been done on facial expression in coatis but given the dramatic eye spots and the proximity that they often interact in, I would be surprised if facial expressions were not important in coati communication.

In the photos I have seen of mountain coatis, their face is entirely dark or gray, with no or faint eye spots, and no white areas on the jaw line. It would be interesting to know if their communication patterns are like those of the *Nasua* species. The white eye spots of white-nosed coatis are present throughout their range; eye spots on brown-nosed coatis are less noticeable in the northwestern part of their range, especially where they overlap with mountain coatis.[18] Newman and colleagues found support for the hypothesis that facial masks in small and medium carnivorans have an aposematic function; warning predators of the danger of attempting an attack.[19,20] This makes the differences in facial patterns in different areas

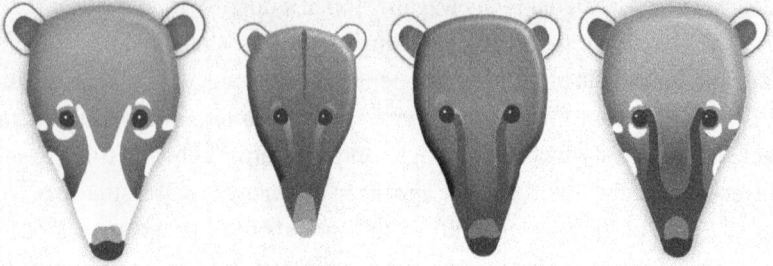

Coati facial patterns north to south. White-nosed coati (far left), western mountain coati (left center), brown-nosed coati from western Columbia (right center) and brown-nosed coati from southern Brazil (right).

even more puzzling, as Andean coatis should be subject to similar predation pressures as more distant populations.

Auditory Signals

Auditory signals have some advantages that visual displays do not. They can travel much farther, and the signaler does not need to be in sight. On the other hand, auditory signals are vulnerable to attenuation in some habitats, so the message may be lost or garbled by the time it reaches the receiver. Like visual signals, auditory signals can carry information about the signalers' identity and emotion.

Much of our understanding of acoustic ecology comes from studies of birds. Birds produce three primary types of vocalizations: mate attraction and territory defense, alarm calls, and contact calls.[21] Across bird species, these types of calls differ in structure depending on the type of message. Calls for mate attraction and territorial defense are highly localizable – the singer wants to advertise its location. It does this by producing complex and distinctive sounds that vary in frequency. These sounds also need to carry a considerable distance, which limits the frequency range, as higher frequency sounds fade quickly with distance and are easily blocked by leaves, trees, and other structures. On the other hand, alarm calls given by a caller that does not want its location known tend to vary little in frequency and are usually simple calls.[1,21] The "seeet" call of passerine birds serves this function, by being able to send an alarm to nearby relatives without giving away information on its own location. In contrast, mobbing calls, issued when smaller birds try to invite a group to attack a bird of prey, are quick, repetitious, and cover a wide frequency range, helping to draw attention to the location of the caller.[1,21]

In contrast to the limited number of visual signals, coatis have a rich acoustic repertoire. They can often be heard before they are seen, as foraging groups produce a variety of sounds – grunts, chirps, chitters, squeals, barks, and growls. Unfortunately, earlier researchers each used their own terminology based on what they thought the vocalizations sounded like, and it is difficult to determine in some cases which vocalization they were describing.

John Kaufmann identified eight different vocalizations among the wild coatis he studied in Panama, with two primary sounds making up most of the vocalizations he heard – a grunt that varied in call rate according to the animal's motivation, and a chitter, which he described as a high-frequency, bird-like call, usually given by juveniles and subadults and apparently related to stress.[22] High intensity chitters graded into squeals. He also described a "chop-chop" vocalization, which was produced by males in a conflict situation, usually with females; a "chuckling" sound issued by coatis when they were mutually grooming or greeting; and an alarm bark when coatis were very alarmed or startled; and whining by juveniles.

In addition to the sounds that Kaufmann documented, Harriet Smith, in her study of captive coatis, documented a "hiss" produced by juveniles when playing, and growls produced by adult females. Both were soft sounds that could only be heard very close to the animals.[23]

Gilbert, in his study of coatis in the southern Huachucas, observed and described numerous vocalizations among the troop members.[24] He recognized three basic sounds – the squeal, the chirp, and the grunt. He also recognized that different sounds could be combined in different ways to produce new sounds, to which he ascribed different meanings. He and his assistants were also limited to sounds they could hear (as were earlier researchers), thus missing any ultrasonic signals.

Few formal studies of coati vocalizations have been done. Two studies focused on captive coatis and focused on only two vocalizations – the chirp and squawk.[25,26] These calls may be related to the grunts and squeals described by Kaufmann and Gilbert. In addition, they recorded the vocalizations, and even used a bat detector (which detects ultrasonic frequencies), so they were able to examine in much greater detail the frequency and detail of the vocalizations.

Intrigued by the results of these studies, I wanted to learn more about the structure and possible function of the various coati vocalizations. So, I combed through my library of video tapes, both those I made during my research as well as some commercial videotapes. I extracted the audio portion of these tapes and converted them to digital files so I could analyze them using spectrograms, which are graphs of frequencies as they vary over time. Information on the intensity of the signal is also included in the graphs. I quickly learned that this was not a very satisfactory way to examine the sounds, as the background noise of the camera plus the ambient sounds and, in some cases the narrator and background music, made visualizing the actual vocalizations difficult. In addition, most of the cameras only recorded up to 15-18 kHz, which meant I could not examine any ultrasonic calls. I bought a portable audio recorder that could record vocalizations up to 48 kHz[27] and in 2011 sent it to the Rolinskis to record their coatis. I knew this might not get the entire range but would be a start. Later, I acquired a recorder that could record vocalizations up to 100 kHz, and I recorded the coatis at the Arizona-Sonora Desert Museum in 2012 and 2013. Once I finished recording the captive animals, I began the long task of listening and visualizing the coati sounds from all the sources. Finally, I started to recognize some patterns, and the detailed spectrogram analysis revealed how complex the coati vocalizations were.

I gleaned at least 22 different vocalizations from the video and audio recordings. I heard two more vocalizations while watching coatis but was unable to get them on tape. From 2010 through 2013, Aline Gasco and colleagues studied vocalizations of brown-nosed coatis in an urban park in Brazil. Gasco identified 15 different vocalizations among coatis, including some combinations of sounds that appeared to create new meanings.[28] Although many of the same calls appeared in the repertoire of both white-nosed and brown-nosed coatis, in some cases they were used differently, and each species had some unique calls. Given the genetic differences, plus being more than 5,600 miles apart, the similarities are striking. In late 2019, Emily Grout began her dissertation work on vocal communication in white-nosed coatis in Panama, and we are collaborating to get a more detailed picture of this important mode of coati communication. I describe the vocalizations in more detail in the following chapters.

Raccoons have been recorded producing up to 13 different vocalizations, most of those used in communication between a mother and her cubs, and in agonistic situations.[29,30] Many of the sounds differed from any sound produced by coatis, although they are described as squeals, chitters, growls, whistles, and barks. None of the vocalizations documented so far exceed 16 kHz, with most vocalizations lower than 6 kHz.[29] Apparently no one has tried to record raccoons using a recorder capable of detecting ultrasonic frequencies, although they can hear well into the ultrasonic range, up to 85 kHz.[4] Likewise, the nine recognized vocalizations of ringtails were less than 7 kHz,[31] although several vocalizations were similar in form to those of coatis. Ringtails produced a "whistle-grunt" that was similar in form to a coati contact call, and captive ringtails were reported to produce this sound in a chorus, indicating that it may function as a contact call. Ringtails also produced chitters, growls, barks and chucks that appear like sounds produced by coatis. Little research has been done on the vocalizations of other species of procyonids.

Olfactory Signals

Although it has long been recognized that scent is important for communication in many animals, the technology to examine odors in detail is still being developed. Mammals deposit scent from urine, feces, and glandular secretions, many of which have a communication function.[1] Other than some studies on scent glands, essentially no research has been done on the topic of chemical signaling in coatis.

Odor has some distinct advantages compared to sight and hearing for communication. Odor can provide detailed information about an animal's identity, gender, health, and reproductive and dominance status.[32] Odor deposited on objects (scent marks) can linger for long periods of time – months in some cases – allowing communication to occur even when the sender is far from the site. Because the scent fades with time, the scent mark also contains information about how long it has been since the sender was in the area, and it also allows a receiver to determine the direction of travel if the scent is fresh enough to detect the difference between one mark and the next. For solitary animals, this allows communication among animals that are normally out of sight and hearing range of each other.[1,33]

Coatis have scent glands in several locations on their bodies. Males have scent glands on their prepuce – the loose skin at the end of the penile shaft, which produce a sebaceous (waxy) material.[34] This waxy material probably carries pheromones, which are chemicals used to communicate with others of the same species.[1] Males frequently scent mark by dragging their abdomen on objects like logs and rock, dribbling a little urine and depositing scent from the glands.[22] They do this year round, but more frequently just before and during the mating season. Females also leave scent, by squatting and dragging their perineum briefly across the ground.[23] No scent glands have been found in the genital area of female coatis, so presumably they are depositing urine. Although it has not been tested, coati urine may contain information on identity, gender, health and reproductive status, and perineal sniffing is common among both male and female coatis, particularly during the mating season.[23]

Coatis are not territorial, and they do not establish latrines other than collections of waste that accumulate near often-used night dens. They are unusual among carnivorans in not having anal pouches that are used for applying scent to feces or for defense.[2,35] They do possess a modified anal gland, but its function is not known. I have collected more than 100 coati feces, and smelled many more, as part of studies on diet (see chapter 13). Coati feces smell like what they have been eating, in contrast to similar gray fox feces, which smell strongly of sweet musk. Both ringtails and raccoons have anal pouches for adding scent to feces; and both create latrines on commonly used travel ways.[36,37] Raccoons also scoot their rump on objects to deposit scent from anal glands.[30]

Coati feet are noticeably stinky. Captive coatis have been observed smearing urine on their feet,[23] and in the film *Nasenbären*, a wild male brown-nosed coati urinates on his rear feet, then shuffles them back and forth, presumably to spread the urine. I have seen coatis released from traps sniff the ground and follow the trail of their troop mates, so at a minimum, scent deposited as coatis are moving allows other coatis to find them.[38] I have also noticed tracks of mountain lions on top of fresh coati tracks, and once observed a mountain lion following a trail of fresh coati diggings, so their predators also appear to be able to follow their scent trails. It is possible that there are scent glands on coati feet, but they have not been described yet. A recent study of glands in the feet of brown bears found 20 different aromatic compounds, and suggested that information about sex, and possibly age was left behind in their tracks.[39] Raccoons also have glands in their feet, however, they appear to help maintain the soft surface of the pads, rather than having any communication function.[40] There are many hypotheses for why animals establish scent marks.[41,42] Little work has been done on this in coatis, but given their social structure and lack of territoriality, I propose that most coati scent marks are for group cohesion and advertising reproductive status.

Is It Language?

Dr. Con Slobodchikoff, in his book, *Chasing Dr. Dolittle: learning the language of animals*, describes language as involving a semantic signal by one animal that is received by another animal within a particular context. Different signals are used in different contexts. Signals and responses to those signals may be learned or instinctive, or both. The important thing is that the signals have specific meanings, and those meanings produce predictable responses. Although many people consider language to be within the domain of humans only, many animals, including coatis, communicate with signals that portray identity, action, and emotion. Dr. Slobodchikoff says,

> *...there is a considerable body of evidence that suggests that animals do have language, and that they can use this language very successfully to warn each other about predators, tell each other about sources of food, tell each other about possible aggression, and signal their intentions to mate. If we approach the premise that animals might have language from an open perspective, then we start seeing all kinds of bits of evidence that it might be true.*[43]

Notes and References

1. Bradbury JW, Vehrencamp SL (2011) Principles of animal communication, 2nd ed. Sinauer Associates, Inc., Sunderland, MA

2. Gompper ME (1995) Nasua narica. Mammalian Species 487:1–10

3. Wiley RH, Richards DG (1978) Physical constraints on acoustic communication in the atmosphere: Implications for the evolution of animal vocalizations. Behavioral Ecology and Sociobiology 3:69–94

4. Peterson EA (1969) Levels of auditory response in fissiped carnivores. Journal of Mammalogy 50:566–578

5. Peterson (1969) reports testing *N. nasua*, but he does not state origin of test subjects, so it is unclear if he was actually testing *N. nasua* or *N. narica*.

6. Heffner HE, Heffner RS (2016) The evolution of mammalian sound localization. *In*: Acoustics Today. http://acousticstoday.org/issues/2016AT/Spring2016/#?page=22. Accessed 24 Apr 2016

7. Arch VS, Narins PM (2008) "Silent" signals: selective forces acting on ultrasonic communication systems in terrestrial vertebrates. Animal Behaviour 76:1423–1428

8. Sunquist M, Sunquist F (2002) Wild cats of the world. University of Chicago Press, Chicago, IL

9. Chausseil M (1992) Evidence for color vision in procyonides: Comparison between diurnal coatis (*Nasua*) and nocturnal kinkajous (*Potos flavus*). Animal Learning and Behavior 20:259-265

10. Personal observation.

11. I recognize that the use of the term "information" here is controversial, with some authors decrying its use given that we have little idea exactly what information is being transferred. In general, they prefer to focus on the actions of the receiver (e.g., Owren et al., (2010)), whereas Dall et al. (2005), Carazo and Font (2010), and Bradbury and Vehrencamp (2011) find that such an approach bypasses the interesting study of the signals themselves.

12. Dall S, Giraldeau L, Olsson O, McNamara J, Stephens D (2005) Information and its use by animals in evolutionary ecology. Trends in Ecology & Evolution 20:187-193

13. Carazo P, Font E (2010) Putting information back into biological communication. Journal of Evolutionary Biology 23:661-669

14. Owren MJ, Rendall D, Ryan MJ (2010) Redefining animal signaling: influence versus information in communication. Biology & Philosophy 25:755-780

15. Walther FR (1984) Communication and expression in hoofed mammals. Indiana University Press, Bloomington, IN

16. Shipp J (2012) Interspecies communication and alarm eavesdropping on multiple birds by a sympatric mammal. Master's Thesis, California State University, Northridge

17. Danchin E, Giraldeau L-A, Valone TJ, Wagner RH (2004) Public information: from nosy neighbors to cultural evolution. Science 305:487-491

18. Personal observation from photos on iNaturalist (https://www.inaturalist.org).

19. Newman C, Buesching CD, Wolff JO (2005) The function of facial masks in "'midguild'" carnivores. Oikos 108:623-633

20. See also Bradbury and Vehrencamp (2011) for an interesting review on aposematic signals.

21. Marler P (1955) Characteristics of some animal calls. Nature 176:6-8

22. Kaufmann JH (1962) Ecology and social behavior of the coati, *Nasua narica*, on Barro Colorado Island, Panama. University of California Publications in Zoology 60:95-222

23. Smith HJ (1980) Behavior of the coati (*Nasua narica*) in captivity. Carnivore 3:88-136

24. Gilbert B (1973) Chulo. A year among the coatimundis. Alfred A. Knopf, New York, NY

25. Compton LA, Clarke JA, Seidensticker J, Ingrisano DR (2001) Acoustic characteristics of white-nosed coati vocalizations: a test of motivation-structural rules. Journal of Mammalogy 82:1054-1058

26. Maurello MA, Clarke JA, Ackley RS (2000) Signature characteristics in contact calls of the white-nosed coati. Journal of Mammalogy 81:415-421

27. Nyquist frequency; the digital sampling rate was twice that.

28. Gasco A, Ferro HF, Monticelli PF (2018) The communicative life of a social carnivore: acoustic repertoire of the ring-tailed coati (*Nasua nasua*). Bioacoustics 28:459–487

29. Sieber OJ (1984) Vocal communication in raccoons (*Procyon lotor*). Behaviour 90:80–113

30. Zeveloff SI (2002) Raccoons: a natural history. Smithsonian Books, Washington, D.C.

31. Willey RB, Richards RE (1981) Vocalizations of the ringtail (*Bassariscus astutus*). The Southwestern Naturalist 26:23–30

32. Scordato ES, Drea CM (2007) Scents and sensibility: information content of olfactory signals in the ringtailed lemur, *Lemur catta*. Animal Behaviour 73:301–314

33. Gorman ML, Trowbridge B (1989) The role of odor in the social lives of carnivores. *In*: Gittleman JL (ed) Carnivore behavior, ecology, and evolution. Cornell University Press, Ithaca, NY, pp 57–88

34. Shannon D, Kitchener AC, Macdonald A (1995) The preputial glands of the coati, *Nasua nasua*. Journal of Zoology, London 236:319–357

35. Gompper ME, Decker DM (1998) Nasua nasua. Mammalian Species 580:1–9

36. Barja I, List R (2006) Faecal marking behaviour in ringtails (*Bassariscus astutus*) during the non-breeding period: spatial characteristics of latrines and single faeces. Chemoecology 16:219–222

37. Hirsch BT, Prange S, Hauver SA, Gehrt SD (2014) Patterns of latrine use by raccoons (*Procyon lotor*) and implication for *Baylisascaris procyonis* transmission. Journal of Wildlife Diseases 50:243–249

38. Also observed by Kaufmann (1962).

39. Sergiel A, Naves J, Kujawski P, et al (2017) Histological, chemical and behavioural evidence of pedal communication in brown bears. Scientific Reports 7:1–10

40. Yasui T, Tsukise A, Meyer W (2004) Histochemical analysis of glycoconjugates in the eccrine glands of the raccoon digital pads. European Journal of Histochemistry 48:393–402

41. Begg CM, Begg KS, Du Toit JT, Mills MGL (2003) Scent-marking behaviour of the honey badger, *Mellivora capensis* (Mustelidae), in the southern Kalahari. Animal Behaviour 66:917–929

42. Muller CA, Manser MB (2008) Scent-marking and intrasexual competition in a cooperative carnivore with low reproductive skew. Ethology 114:174–185

43. Slobodchikoff C (2012) Chasing Doctor Dolittle: learning the language of animals. St. Martin's Press, New York, NY

10

Keeping the Group Together

Most coati displays and other behaviors serve to keep the group together, by letting individuals stay in contact, by communicating intent, and by diffusing potentially aggressive behavior. These affiliative behaviors, which may consist of individual behaviors or several given simultaneously, occur in a variety of contexts, including foraging, greeting, allogrooming, and play behavior.

Foraging

As a troop of coatis forages through the leaf litter, they often carry their tails upright; furry semaphores which, combined with movement of the animals as they forage, increases their visibility in the forest. They also issue contact calls as they are foraging, described as "grunts" or "chirps" by previous researchers. My detailed examination, also found by Gasco and colleagues,[1] revealed a complex foraging call that may convey a large amount of information. These contact calls, called chirps, are produced when the coatis are resting, foraging, or moving.

Chirps usually consisted of two or three (up to eight) ascending notes given together, one usually starting at around 6 kHz, and another starting at around 13 kHz and ascending to 16 or 17 kHz. Sometimes the higher note was split into several components, and the highest component may extend up to 24 kHz. The higher notes were not harmonics, rather they were biphonation or polyphonation.[2] The calls were quick, lasting less

Typical tail-raised posture of foraging white-nosed coatis. Guatemala, 2007.

than 0.2 seconds. Chirps differed in their length and how fast they ascended, but they always ascended in frequency. The call rate of the chirps appeared to reflect the arousal level of the coatis. They increased their rate of chirping when meeting new coatis, if they found some interesting food, after aggressive interactions, when rejoining the troop after separation, or when it was time for the entire group to move.[3] Chirps are considered the coatis primary contact call.[1,4]

Coatis frequently followed the chirp with a quick, low-frequency grunt. The grunt is 0.04 to 0.6 kHz – near the lower level of human hearing, but sometimes loud enough to be the most audible part of the whole call. Grunts are rarely given on their own but are usually tacked on to the end of a chirp or given simultaneously with a bark (see Chapter 12). They vary from a soft, quick "unh" to a loud grunt or pop. Grunts may function by adding emphasis to the accompanying call. Gasco and colleagues described a "subsound," which appears to be the same call, although they described its use in more aggressive contexts.[1] Some primates, such as redfronted lemurs from Madagascar, use grunts to coordinate movements in a group.[5]

In addition, some versions of the contact call of white-nosed coatis include a very quick broadband click. This sound rarely appeared by itself but might be included with chirps, grunts, or chirp-grunts (in which case, the call was a chirp-click-grunt). The click ranges from about 0.5 kHz to

7.5 kHz. It is often given by adult females in a troop or with their young-sters when they are moving and seems to function as sort of a "come-along" or "let's go" type of call. In my studies of white-nosed coatis, wild coatis added clicks to their contact calls 84% of the time, whereas captive coatis added clicks only 4% of the time.[6] The captive coatis were generally loafing when being recorded, while the wild coatis were usually active and foraging. Gasco and colleagues identified a low-frequency, harmonic squeak-like sound in brown-nosed coatis that appeared just before the grunt, which they labeled the "coo."[1] The coo was issued when coatis were foraging and disputing food morsels.

Earlier researchers have considered the contact call, in all its forms, as one call, variously called a "grunt" or "chirp." After detailed examination, I discovered that components may be given on their own or combined with others, so I think it is important to treat them separately. The entire chirp-click-grunt call is less than one second in duration, making it diffi-cult for humans to hear the separate components. Coatis obviously can process sound more quickly than humans can. Chirp-click-grunts differed in the amount of energy devoted to different components of the signal, so that an individual may emphasize the chirp in one call, and a few seconds later emphasize the grunt. As Kaufmann noted, *"These basic vocal patterns could be subdivided endlessly according to small morphological differences."*[3] Kaufmann also described the use of contact calls not only for foraging, but also for reuniting the group, especially in response to juvenile chitters. Both white-nosed and brown-nosed coatis produced similar chirps and chirp-grunts but differed in additional components like the click and coo.

The chirp, as a contact call, not only lets group members know where other group members are, but also contains information about individual identity.[1,4,7] So, when a group of coatis was out foraging, various adults pe-riodically issue chirps, letting other group members know who was where. Gasco and colleagues suggested these individually distinctive calls could facilitate group cohesion.[1] In the small captive groups that Maurello and Compton studied, there appeared to be what they termed a "control ani-mal" who called first, followed by the other animals.[4,7] I believe that I saw similar phenomena among the wild coatis, but I was not able to gather enough data to verify. I was able to distinguish among a few of the female coatis by their chirp-grunts.

Research on vocalizations in meerkats may provide some insight here. Meerkats are small group-living carnivorans that live in the Kalahari

Desert of southern Africa. They are diurnal and feed on insects, much like coatis. However, they live in open habitats, whereas coatis live in the forest. As the meerkats are foraging for insects, they periodically issue vocalizations. They use three types of vocalizations; a close call issued by all group members which appears to provide information on who is where; a lead call, issued by animals moving fast in a single direction; and a moving call that stimulates the group to move or keep moving.[8] Townsend and colleagues described meerkat close calls as a form of "all's well" vocalization, which allowed group members to coordinate vigilance and relay information about the lack of current predator threats.[9] It's very possible that the chirp, chirp-grunt, and other variations, are separate calls with different meanings, some of which may be analogous to the close calls of meerkats. But it will take more observations and recordings, as well as playback experiments, to reveal the meanings of the different components. Unfortunately, coatis are not as easy to work with as meerkats.

Playbacks of contact calls of resident coatis to the coatis at the Arizona-Sonora Desert Museum elicited no visible responses. However, given the quiet nature of the calls, it is possible that the coatis did not hear the broadcast calls, or that their responses were soft vocalizations and not overtly physical. Playback of a chirp-click-grunt sequence from a wild female with her young kits caused the Desert Museum coatis to move away from the sound. When I went back to the computer and re-examined the sounds that I had broadcast to the coatis, I saw why they moved away: in addition to the chirp-click-grunts, there were a few alarm barks (see Chapter 12) interspersed with the sequence, and the grunts were very loud. This suggests that coatis maybe able to encode some component of "danger." Gasco and colleagues identified a syntax in coati vocalizations, where a chirp-coo-grunt is a foraging call and a chirp-grunt-bark is an alarm call or threat.[1]

Coatis share their forested, often jungle, environments with many species of birds. They are active during the day, at the same time birds are calling. Most birds in the American tropics call between 2 and 12 kHz.[10] The upper and lower portions of the contact call are therefore above and below most of the avian noise of the jungle. This contrasts with the friendly vocalizations of raccoons, which occur in the 0.5 to 8 kHz range; the same as birds.[11] But raccoons are active during nighttime, when fewer birds are calling.

Chitters of coatis are also heard from a foraging troop. Chitters are high-frequency, bird-like sounds. On a spectrogram, chitters look like a row of upside-down "U"s, as each call ascends to a peak, and then descends to near its starting frequency. Each call is quick, only lasting 0.07 to 0.13 seconds, in the range of 4-9 kHz. They are given in series, of 4-10 per second. They are also given by juveniles in play or when being groomed; and are given by nestlings when just a few days old. Coatis, especially juveniles, separated from the troop issue chitters, and troop members usually respond by approaching the animal that is chittering.[3,12] Chitters appear to indicate mild distress or excitement. Chitters resemble bird calls (for example, Bridled Titmouse), both audibly and in the spectrograms, perhaps helping minimize detection by predators. They occasionally are heard from adults while grooming.

Adult and juvenile white-nosed coatis also make squeaking sounds as they are foraging, moving, and playing. Squeaks are a fast, harmonic sound that ranges from 2-15 kHz. Although to human ears, they could be confused with a chirp, they looked quite different on a spectrogram. Squeaks appear to reflect excitement or surprise. Brown-nosed coatis use a similar vocalization when playing.[1] Captive coatis issue whimper and twitter vocalizations when anticipating food, or release from their night den. These also appear to be indications of excitement, but both are softer vocalizations and not recorded often. Whimpers were low frequency calls (0.2-0.4 kHz), less than 0.5 seconds long, and issued in irregular clusters. Twitters were high frequency (7.5–11 kHz), and also issued in irregular clusters, meaning that the coatis would repeatedly twitter or whimper for seconds to minutes at a time and then switch to another call or go quiet. They appeared to elicit no responses from the other coatis, but just appeared to be measures of excitement or anticipation. As such, these sounds may not directly function to keep the group together but may provide public information about the arousal level of individuals in the group.

Grooming and Allogrooming

Touch is especially important for communication among coatis. In addition to self-grooming, they groom each other, sleep on top of each other, and have noisy greeting ceremonies that involve a lot of touching and licking. They often rest side by side, in contact and frequently touch and nuzzle each other.[3,12,13]

Coatis groom themselves using both their teeth and claws, picking off ectoparasites with their tiny, comb-like, front incisors, and scratching vigorously with alternating front feet. They will also scratch their face, neck, and shoulders with their back feet, much like a dog or cat. Coatis groom each other by using their front teeth to comb each other's fur, in quick nibbles. Mutual grooming is often initiated when one animal begins self-grooming, and others approach and join in. The initiator may or may not reciprocate.[3,12] Grooming sessions sometimes start with the coatis sitting next to each other, facing opposite directions. They begin with grooming each other's shoulders, neck, and head. Adult females do most of the grooming, usually directed to their offspring, but also to other adult females in the group, as well as any adult males associated with the group.[14]

Harriet Smith differentiated inhibited biting, when coatis approached each other and quickly nibbled the forequarters, neck, and head, from true mutual grooming, which is more prolonged and involved more regions of the body.[12] Although both likely have a role in social bonding, only the latter would have much of a role in reducing ectoparasites. As mentioned in an earlier chapter, coatis that are the targets of more mutual grooming, including other adult females and juveniles, have fewer ticks and mites than animals that are groomed less frequently, such as yearlings and adult males. Inhibited biting was common just before coatis went to sleep or upon first waking and was sometimes followed by mutual grooming. It often involved more than two animals. Inhibited biting was also used by an animal attempting to break up a fight between two other coatis and seemed to help reduce tension.[12]

Mutual grooming was most common before and during the breeding season and may be one way for females to monitor and synchronize each other's reproductive condition. Coatis appeared to have remarkable powers of recuperation, attributed to mutual grooming, which included licking of wounds.[3,15] Perhaps due to the lower levels of parasites in the northern part of their range, I seldom observed mutual grooming. Mutual grooming, including the inhibited biting described above, has been shown to have important social bonding functions in a number of species, including social meerkats.[16]

Kaufmann also described "head jerking" by juveniles, which appeared to stimulate adults to groom the juveniles.[3] Smith mentioned head jerking by adults,[12] and I've also observed head jerking or head tossing by adults, particularly during nose-up displays. I think that there is some meaning to this behavior, but no one has looked at it in detail yet. It may be that quick movements help focus visual attention on the animal doing the displaying.

When mutual grooming, white-nosed coatis sometimes make an excited, variable riot of sounds, which Kaufmann termed "chuckles." They consist of a variety of chirps, chitters, squeaks, clucks, giggles, and other noises, all given rapidly and showing different forms than when given in other situations. Their vocalizations cover a broad frequency range, with the highest frequencies approaching 30 kHz. Some of the sounds are structurally like chirps, although entirely new sounds are also present. Often the same syllables are repeated for several seconds, and then a new sound begins. Coatis that have been separated for various periods of time may also make this vocalization, with the amount of the vocalizing depending on how excited the coatis are to see each other. This vocalization may be a form of greeting song and may help diffuse tension when one coati approaches another. When two friendly coatis greet, they grasp each other's head, nuzzle and gently bite and lick each other, with both animals making this variety of sounds, although it was not possible in the recordings to determine which coatis were making which sound. Captive coatis may also greet their handlers with this vocal and tactile display. It is given by all ages and sexes, but most often by adults during the mating season.[3] Brown-nosed coatis issued chittering vocalizations when grooming and allogrooming, but apparently not the excited chuckles.[1]

Contact calls, squeaks, chitters, and chuckles comprise a suite of affiliative calls. Among both captive and wild white-nosed coatis, affiliative calls made up at least 80% of the recorded repertoire,[17] indicating that these calls may indeed be associated with group structure and cohesion.

Coatis also add scent to their bodies and tails. Gompper and Hoylman described coatis adding the resin of *Trattinnickia aspera* to their tails and bodies in Panama, and suggested that they might be self-medicating to reduce parasites.[18] *Trattinnickia aspera* only occurs in Panama, but the genus *Trattinnickia* occurs from Costa Rica to Brazil. Numerous other members of the family Burseraceae (which also includes the plants that produce frankincense and myrrh) are found throughout coati range except the very northern extent, and many are known for their medicinal properties, including repelling insect pests.[19] Both males and females applied the *Trattinnickia* resin to their bodies, but there were significant differences in the ectoparasite loads between group living and solo coatis, indicating that allogrooming may have been more important than the *Trattinnickia* in reducing parasites.

Coatis appear to be fascinated by aromatic compounds (*Trattinnickia* contains terpentine-, camphor-, and menthol-like compounds).[18,20] Interestingly enough, although coatis were observed smearing *Trattinnickia* resin all over their bodies, other aromatic compounds are applied just to the tail, suggesting that there may be more than one thing going on. My conversations with the Rolinskis revealed that coatis like to rub anything aromatic on their tails, especially if it is a novel odor. Expensive perfume is a favorite, but only if they have not been around it in a while. I saw this firsthand on a visit to the Rolinskis and noticed that the coatis used a very stereotyped set of behaviors to apply the scent to their tails. First, they rubbed the area under their nose in the scent. On examination, I found that the area between the bottom of the coatis' nose and its upper lip is made of a spongy, almost naked area of flesh. They then use their claws to separate the fur on the underside of their tail and rub the underside of their nose on the exposed skin. They repeat this action several times. They did not use their forefeet to collect and apply the scent, as Gompper and Hoylman described coatis doing to distribute the *Trattinnickia* resin.

In Brazil, brown-nosed coatis were seen applying laundry soap and cleaning substances to their bodies. They would usually apply the soap to themselves, using their noses and front feet, but also rubbed the substances on other coatis, sometimes having group anointing sessions. Most of the substances were applied to the genital region and tail but also included other parts of their bodies. Young coatis appeared to learn this behavior from older coatis. The purpose of the anointing was not known, but may have had some parasite-repellant function.[21] Similar behavior was observed among captive white-nosed coatis.[20] Brown-nosed coatis have also been observed rubbing millipedes in their fur; millepede secretions contain tick-repellent compounds.[21]

Remote camera photos taken as part of research conducted by the Borderlands Jaguar Detection Project in southern Arizona show groups of coatis smearing the lure used to attract carnivores to the camera sites on their tails.[22] The lure used at the time was a rotten, skunk-based lure. It is probable that adding odor to their tails is another way for a coati group to stay together, but in this case, they were adding an odor known to attract canine and feline predators. Mountain lions in the area did kill and feed on coatis,[23,24] but it is not known if the lure added to their tails made them more vulnerable to predation, and this lure was not used in the Huachucas when I documented high levels of mountain lion predation on coatis there.

Whether coatis just apply some materials to their bodies and others just to their tails is not known. Coatis might have learned about *Trattinnickia* and soap anointing through cultural transmission.[18,21] It is possible that coatis apply novel scents on their tails as a group, to create a "group smell." Solo males also apply these compounds to their tails in similar fashion as group members do. Perhaps this helps groups to distinguish local males from foreigners. Scent is most likely the coatis' most important means of communication and collecting information from the environment, yet it is the one we know the least about.

Coatis are not the only mammals that apply scent to their tails. Ringtailed lemurs are social, diurnal primates that live in female-dominated groups in Madagascar. They have long, ringed tails, to which the males apply scent from glands on their wrists. The males then wave their tails towards opponents in "stink fights" and toward prospective mates in "stink flirts."[25,26] They also use scent glands on their chest, shoulders, and scrotum to mark their territories, as well as advertise reproductive condition and genetic fitness.[25,27,28] Although ecologically similar to coatis, olfactory communication in ringtailed lemurs appears to be more complex.

Play Behavior

Coatis of all ages play, although most play is by juveniles. Before they are six months old, the youngsters chase each other up and down trees, periodically stopping to spar and wrestle, while the adults and subadults are busy foraging. Although coatis are more closely related to dogs, they play more like cats. Play consists of wrestling, biting, sparring, chasing, and lunging. A coati trying to start a play bout typically stands on its rear legs with its front legs spread wide, and its mouth open in a "play face." Coatis also show a desire to play by bouncing sideways, much like kittens. They issue a variety of chitters, squeaks, hisses, and buzzes during play.[1,3,12]

In most mammals, play behavior serves a training function, allowing young animals to practice adult behaviors in a safe environment. They learn the proper social contexts for different behaviors, and how much they can get away with without hurting another or generating an attack. At the same time, they develop muscles and physical skills that they need to move through their environment and escape from predators.[29-31] Play

Juvenile coatis in play. Left, stance of a juvenile coati inviting another to play, right, wrestling. Guatemala, 2007.

may also help develop and maintain social bonds, which may be particularly important for social animals such as coatis, and may develop the behavioral flexibility to deal with unknown situations.[32] Even adult males, normally quite nasty to each other in the wild, will play with subadults ocassionally,[1,32] and with other adult males in captivity.[33] No detailed studies of play in coatis have been conducted, in which individual behaviors performed by and to individual participants have been recorded. These types of studies are needed to examine the development of behavior[30] and the importance of play in developing or maintaining social relationships.[29,34]

Other Grouping Behaviors

Michael Sutor placed a microphone in the den of a female coati with newborn young during the filming of *Nasenbären*. The baby coatis made soft whines, mews, and purrs while nursing. These sounds were made in similar contexts as mewing in newborn puppies and kittens, and probably indicated mild distress or contentment. They also made soft whistling sounds when interacting with each other. No microphones capable of recording ultrasounds have been used with newborn coatis, so it is unknown if ultrasounds were produced by the mothers or offspring.

In addition, I discovered from watching hours of videos over and over, that they may use their tails for comforting each other. I first noticed this in a video taken in Ramsey Canyon of a foraging group. One subadult female was foraging in the leaf litter when she was approached from behind by a 4-month-old juvenile. She lunged at the juvenile, who let out some loud chitters as she knocked it over. She immediately stopped when an adult female approached, and she turned back to the area where she had been digging, followed by the juvenile, who stayed behind her. The subadult then draped her tail over the back of the juvenile, and they foraged in that position for several more minutes.

About 10 minutes later, as the coatis were foraging, two adult females jumped into a tree after being startled by something on the far side of the creek. The two females then jumped to the ground and stood quietly while looking around. There was a juvenile with them, and one of the females draped her tail over its back. A few moments later, a person could be seen walking on the far side of the creek. Only when the person was in view did one of the females begin issuing an alarm bark. I have termed this behavior a "tail-hug."

I saw similar behavior between two adult female coatis in Guatemala. They had sensed something in the jungle near where they were feeding and jumped up into a tree to investigate, where I observed them standing side by side on a branch. As they tried to discern what was approaching (an adult male), they were waving their tails back and forth, and would pause for a second when their tails were over each other's back. It was like the tail-hug I saw in Arizona, but less deliberate, and may have been an incidental part of the tail waving (see photo in Chapter 12).

This behavior has not been described in coatis before. Although my sample size is small, the behavior appeared very deliberate, and the coatis kept their tails in contact with the other animals for at least 15 seconds. The behavior appeared to be an attempt at comforting or consoling a juvenile or other troop member. The first instance I described may have been a form of post-conflict reconciliation,[35] but the others appeared to be more of a comforting nature. Scientists have long been reluctant to ascribe emotions such as empathy to other animals, but more and more research reveals that animals including dogs, primates, and rodents care and show concern for others of their species. Coatis exhibit a number of grouping behaviors that indicate concern for their troop-mates, and I think the tail-hug is one such behavior.[36,37]

Kaufmann also described coatis using their tails in social interactions.[3] During the mating season, adult males approaching troops would sit or lie near them and orient their tails toward the group. He also saw this behavior by adult females approaching adult males. This "tail-to" appears to be a friendly gesture as if the approaching animal is showing that it means no

Adult female with her tail draped over the back of a juvenile. The coatis are "freezing" while trying to determine the source of an alarm sent out by other coatis. Ramsey Canyon, 1994. Photo extracted from videotape.

harm. I did not detect this behavior, nor has anyone else described it, but it appears to be subtle and perhaps easily missed.

Notes and References

1. Gasco A, Ferro HF, Monticelli PF (2018) The communicative life of a social carnivore: acoustic repertoire of the ring-tailed coati (*Nasua nasua*). Bioacoustics 28:459–487

2. The production of multiple sounds simultaneously, usually achieved through modification of the vocal folds.

3. Kaufmann JH (1962) Ecology and social behavior of the coati, *Nasua narica*, on Barro Colorado Island, Panama. University of California Publications in Zoology 60:95–222

4. Maurello MA, Clarke JA, Ackley RS (2000) Signature characteristics in contact calls of the white-nosed coati. Journal of Mammalogy 81:415–421

5. Sperber AL, Werner LM, Kappeler PM, Fichtel C (2017) Grunt to go—Vocal coordination of group movements in redfronted lemurs. Ethology 123:894–905

6. N = 115 contact calls of wild coatis and n = 1009 for captive coatis.

7. Compton LA, Clarke JA, Seidensticker J, Ingrisano DR (2001) Acoustic characteristics of white-nosed coati vocalizations: a test of motivation-structural rules. Journal of Mammalogy 82:1054–1058

8. Bousquet CAH, Sumpter DJT, Manser MB (2011) Moving calls: a vocal mechanism underlying quorum decisions in cohesive groups. Proceedings of the Royal Society of London B 278:1482–1488

9. Townsend SW, Zöttl M, Manser MB (2011) All clear? Meerkats attend to contextual information in close calls to coordinate vigilance. Behavioral Ecology and Sociobiology 65:1927–1934

10. Bradbury JW, Vehrencamp SL (2011) Principles of animal communication, 2nd ed. Sinauer Associates, Inc., Sunderland, MA

11. Sieber OJ (1984) Vocal communication in raccoons (*Procyon lotor*). Behaviour 90:80–113

12. Smith HJ (1980) Behavior of the coati (*Nasua narica*) in captivity. Carnivore 3:88–136

13. Trudgian MA (1995) A study of captive brown-nosed coatis, *Nasua nasua*: an ethogram and contact call analysis. Master's Thesis, University of Northern Colorado, Greeley

14. Hirsch BT, Stanton MA, Maldonado JE (2012) Kinship shapes affiliative social networks but not aggression in ring-tailed coatis. PLoS ONE 7:e37301

15. Personal observation.

16. Kutsukake N, Clutton-Brock TH (2006) Social functions of allogrooming in cooperatively breeding meerkats. Animal Behaviour 72:1059–1068

17. Affiliative calls: wild coatis, 80% of 200 calls; captive coatis, 95% of 1132 calls.

18. Gompper ME, Hoylman AM (1993) Grooming with *Trattinnickia* resin: possible pharmaceutical plant use by coatis in Panama. Journal of Tropical Ecology 9:533–540

19. https://en.wikipedia.org/wiki/Burseraceae (accessed 9/13/2012).

20. Kaufmann JH, Kaufmann A (1963) Some comments on the relationship between field and laboratory studies of behaviour, with special reference to coatis. Animal Behaviour 11:464–469

21. Gasco ADC, Pérez-Acosta AM, Monticelli PF (2016) Ring-tailed coatis anointing with soap: a new variation of self-medication culture? International Journal of Comparative Psychology 29:1–12

22. Emil McCain and Jack Childs, personal communication.

23. McCain EB, Childs JL (2008) Evidence of resident jaguars (*Panthera onca*) in the Southwestern United States and the implications for conservation. Journal of Mammalogy 89:1–10

24. Emil McCain, personal communication.

25. Kappeler PM (1990) Social status and scent-marking behaviour in *Lemur catta*. Animal Behaviour 40:774–776

26. Sauther ML, Sussman RW, Gould L (1999) The socioecology of the ringtailed lemur: thirty-five years of research. Evolutionary Anthropology 8:120–132

27. Charpentier MJE, Boulet M, Drea CM (2008) Smelling right: the scent of male lemurs advertises genetic quality and relatedness. Molecular Ecology 17:3225–3233

28. Scordato ES, Drea CM (2007) Scents and sensibility: information content of olfactory signals in the ringtailed lemur, *Lemur catta*. Animal Behaviour 73:301–314

29. Fagen RM (1981) Animal play behavior. Oxford University Press, New York, NY

30. Hass CC, Jenni DA (1993) Social play among juvenile bighorn sheep: structure, development, and relationship to adult behavior. Ethology 93:105–116

31. Bekoff M (1978) Social play: structure, function, and the evolution of a cooperative social behavior. *In*: Burghardt G (ed) The development of behavior: comparative and evolutionary aspects. Garland Press, New York, NY, pp 367–383

32. Logan CJ, Longino JT (2013) Adult male coatis play with a band of juveniles. Brazilian Journal of Biology 73:353–355

33. Stan and Linda Rolinski, personal communication.

34. Bekoff M, Byers JA (1998) Animal play: evolutionary, comparative and ecological Perspectives. Cambridge University Press, Cambridge, UK

35. Sima MJ (2018) Ontogeny of cognition and communication in Corvids. Ph.D. Dissertation, Universitat Konstanz, Konstanz, GER

36. Burkett JP, Andari E, Johnson ZV, Curry DC, Waal FBM de, Young LJ (2016) Oxytocin-dependent consolation behavior in rodents. Science 351:375–378

37. Safina C (2016) Beyond words: What animals think and feel. Picador, New York, NY

11

Settling Disputes

Although all-out fights among coatis are rare, they do occur. However, most altercations are momentary squabbles over food or space, and these usually involve a lot of vocalizing. Although most communication among coatis is friendly, they can also be quite testy. They do not share food and they can be picky about who gets to hang out in the troop. Baby coatis up to nine months old can get away with almost anything, backed up by their mothers, whereas yearlings are at the bottom of the hierarchy and are picked on by all.[1,2] Although adult males are larger than adult females, females will often gang up on males to drive them from an area. Aggressive and defensive behaviors include a few visual displays, several vocalizations, and physical contact.

Agonistic Visual Displays

The primary agonistic visual displays are termed "nose-up" and "head-down."[2,3] Both of these are graded signals, meaning that they occur in degrees according to the emotion and motivation of the animal. In other words, a mildly perturbed coati might direct a "nose-up" to another coati, then immediately go back to what it was doing, whereas a coati in attack mode might lunge toward another coati while giving the "nose-up" display and vocalizing loudly.

The "nose-up" display consists of the coati lifting the end of its nose. It is a display of weapons (teeth), and in its extreme form, the mouth is open and teeth are bared. A coati showing this display is ready to attack. At its mildest form, the nose is lifted but the mouth is closed, and teeth are not bared. This is just a warning. Body posture also accompanies the level of threat: in attack mode, the coati is standing on all four feet, with its tail straight behind it, and trying to look as large as possible. In its mildest form, there may be no postural changes, but the coati may turn toward another and lift its nose as a mild warning.

Coatis also lift their nose to sniff the air, and they lift their nose to move it out of the way when they are drinking. A coati that lifts its nose as part of a nose-up display often vocalizes at the same time and has its ears back. In contrast, a coati that lifts its nose to sniff the air is quiet and has its ears forward. Coatis also use the nose-up display when playing but indicate lack of aggressive intent with other behaviors (see previous chapter).

The "head-down" signal is just the opposite of "nose-up," in both posture and meaning. It is a submissive signal, so coatis intentionally try to hide their dental weapons by hunching over with their nose pointed toward the ground. They often turn their side to their attacker and move away. This signal is also graded, so in its mildest form the coati may simply look down and slightly deflect its head. In its most extreme form, the coati lays flat on the ground, on its belly, with its head and neck flat against the ground and may even cover its face with its front paws.[2] Intermediate stages often reflect a conflict between fight and flee, which the coati expresses by alternating between nose-up and head-down, and yawning.[2] In some cases, animals yawn as a display of weapons; in this case, yawning appears to be a displacement activity or calming signal,[4] another indication of emotional conflict.

These two behaviors illustrate the principle of antithesis in animal signals, first described by Charles Darwin.[5] As Darwin wrote, opposite emotions are reflected in opposite behaviors. Threatening animals try to look as large as they can, whereas submissive or frightened animals try to look small and non-threatening. For example, dominant or threatening animals show or display weapons such as antlers or horns, by standing taller, or pointing them at opponents.[6]

Agonistic Vocal Displays

Although agonistic vocalizations make up less than 10% of the repertoire,[7] they are louder and more intense and seem more memorable by

Examples of nose-up displays. Left, mild nose-up among playing white-nosed coatis. Right, extreme nose-up among fighting male brown-nosed coatis (photo by Ben Hirsch).

coati watchers. Most squabbles occur while foraging, as a coati attempts to guard whatever it is eating. Coatis frequently lunge at an opponent as a form of bluff-charge. The lunge is accompanied by a nose-up display and loud chittering, and they may swat or bite if the other coati does not move away. Loud chitters are structurally like quiet chitters but start at a slightly lower frequency and include several harmonics, extending the vocalization to 20 kHz or higher. Sometimes they take on somewhat of a rolling quality. Loud chitters appear to be a mobbing call and are issued by juveniles that are attacking adults or subadults and by adults as a "call-to-arms" when attacking a predator or a male approaching the group. In this way, loud chitters could function in bringing coalition support. Female coatis whose babies are being attacked in their nests will chitter loudly, as will the babies as a cry for help. In groups, multiple animals will start loud chittering, bringing even more attention to the situation. This vocalization was used by both white-nosed and brown-nosed coatis, although Gasco and colleagues referred to it as a trill or twitter.[8,9]

Several species of primates use screams to call in group members for assistance in agonistic situations,[10] and the loud chitters of coatis may be analogous. Primate screams are often context-specific and vary with age and ranking of the participants in the encounter. It is not known what kind of information is transmitted in coati chitters, but they vary in intensity, which may indicate that they are communicating urgency, rather than context. Playback of loud chitters from wild coatis to the coatis at the Arizona-Sonora Desert Museum caused several animals to quickly move toward the source of the sound.

White-nosed coatis issue squeals when frustrated or outraged. They are very loud, descending calls, although they could also appear as a noisy combination of squeals and squawks (see below). Loud chittering grades into an unbroken squeal, which is accompanied by a nose-up display and is often followed by an attack.[2] The calls are lower in frequency than loud chitters, with the lowest portion of the call between 2 and 5 kHz. Squeals include several harmonic bands extending to 40 kHz or more, although most of the energy is below 15 kHz. This vocalization was also graded, with more intense situations or more outraged coatis issuing more intense squeals. Playback of squeals from their own group at the Desert Museum elicited an approach by an older male, who was known to routinely break up fights among the other coatis. A squawk is like a squeal but includes a lower throaty component and tends to be issued more by adult females, but in the same circumstances as squeals. Squawks and squeals were not reported in brown-nosed coatis.[8]

Rarely, coatis in traps and involved in fights were heard giving very loud, low frequency roars. The calls were noisy and ranged from 0.1 to 10 kHz. Roars also accompanied squeals and were occasionally heard with alarm calls (chapter 12). The calls were issued by coatis that appear to feel threatened, such as when a dog treed a coati, or female coatis were defending their newborns against attack. The harshness of the call left little doubt as to its meaning. Both white-nosed and brown-nosed coatis used this vocalization.

Coatis also issue a chop-chop vocalization, which is a rapid teeth chatter. It is primarily issued by males during the breeding season in white-nosed coatis (see chapter 15 for an example) and is hypothesized to reflect conflict between attack and escape.[2,3] Juvenile and female brown-nosed coatis also issued this vocalization when restrained by hand or in traps, and Gasco and colleagues described its use *"in situations of high excitement inherent to alarm, threatening and distress situations."*[8] It often accompanies other calls, such as barks and roars.

Fighting

Although minor squabbles are common, actual fights among individuals are rare. Minor fights or squabbles occur over disputed access to food or resting sites.[2,8] Kaufmann described this as "inhibited fighting," which included swatting the partner's head with one or both forepaws, inhibited biting of the partner's head, and sometimes mututal sparring with the mouth open but not the use of paws.[2] It was often accompanied by squawks or loud chitters.

However, when serious fights occur, they can be very violent and bloody. The most vicious fights occur between males during the mating season, and by the end of the season, males end up with ripped flanks, torn lips and ears, and could often be seen limping.[2,9] One male even lost his lower canines in fights. Females are also injured in fights during the mating season, presumably from fights with males, and end up with ripped skin and puncture wounds. Fights are accompanied by a lot of loud squealing and roaring. Most of the damage is caused by sharp canine teeth, but claws are also used.

One male that Jason and I followed quite closely during the start of the study was M14. He was somewhat habituated to people and could often be seen at the picnic areas in Garden Canyon. In April 1996, hikers in Scheelite Canyon observed M14 in a fight with another coati, and described it in the trail register (my comments in brackets):

Halfway between 3/8 and 4/8 [mile markers]. Heard raccoon-like screeching behind us as we descended trail. Ran back where 2 oaks frame path. Saw 2 coati tails going over edge of path. With much scuffling and screeching, ended up behind rocks in streambed w/one subduing other...Resumed fighting viciously for about 2 minutes. Near exhaustion, they broke apart panting audibly – sides heaving, mouths open dripping blood. Blue collar [M14], who stayed, had large gash on back and cut on left foreleg. Large one who wandered away had gash on left side. Whole event took about 5 minutes.[11]

We saw M14 two days later at the trailhead. He moved gingerly, limping on his left front leg. His left eye was swollen shut, and he had gaping wounds across his shoulders and flanks. He appeared to make a full recovery, but he died of unknown causes seven months later, and I could not determine if any of his old injuries contributed to his demise.

Fights between males also occurred outside of the mating season. During late August 1996, I was radio-tracking coatis in Huachuca Canyon when I found M77, an adult male, up in a chokecherry tree eating the ripe berries. As I quietly watched him, I detected another adult male, M42, from his telemetry signal. The signal got louder and louder, indicating that he was getting closer. I was a little surprised, as we seldom located M42 this far down the canyon. I kept an eye on M77 in the chokecherry tree and followed M42s approach from the beeps of the telemetry receiver. When M42's signal indicated he was close, M77 dashed out of the tree and

M14 following a fight with another male during the breeding season.

I soon heard a loud coati scuffle. They were across a dry stream bed, and the vegetation was too thick to see anything. I tried to sneak my way closer to see what was going on, but I could not approach without crashing through the brush, making a lot of noise and scaring them off. So, I listened in frustration to the loud growls, roars, and squeals, wanting desperately to see what was happening. They separated after several minutes, M42 heading back up the canyon, and M77 heading down canyon. But most of battles, and by far the most vicious ones, occurred during the mating season.

Gang fights (coalitions) occurred when a male tried to join a troop or when youngsters tried to chase a male away. Most of these attacks occurred against adult or subadult males and involved a lot of loud chittering and squealing as more and more animals gathered to attack the offending animal. Although he might initially try to defend himself by swatting with his forepaws to fend off the attackers,[2] the usual result was for the male to flee; his attackers seldom pursued.

Notes and References

1. Hirsch BT (2007) Spoiled brats: is extreme juvenile agonism in ring-tailed coatis (*Nasua nasua*) dominance or tolerated aggression? Ethology 113:446–456

2. Kaufmann JH (1962) Ecology and social behavior of the coati, *Nasua narica*, on Barro Colorado Island, Panama. University of California Publications in Zoology 60:95–222

3. Smith HJ (1980) Behavior of the coati (*Nasua narica*) in captivity. Carnivore 3:88–136

4. Rugaas T (2005) On talking terms with dogs: calming signals, 2nd edition. Dogwise Publishing, Wenatchee, WA

5. Darwin C (1872) The expression of the emotions in man and animals. D. Appleton, New York, NY

6. Walther FR (1984) Communication and expression in hoofed mammals. Indiana University Press, Bloomington, IN

7. Agonistic calls: wild coatis, 8% of 157 calls; captive coatis, 5% of 1132 calls.

8. Gasco A, Ferro HF, Monticelli PF (2018) The communicative life of a social carnivore: acoustic repertoire of the ring-tailed coati (*Nasua nasua*). Bioacoustics 28:459–487

9. Personal observation.

10. Gouzoules H, Gouzoules S (1989) Design features and developmental modification of pigtail macaque, *Macaca nemestrina*, agonistic screams. Animal Behaviour 37:383–401

11. Unknown hikers, Scheelite Canyon trail register, April 2, 1996. Courtesy of Fort Huachuca Wildlife Office.

12

Sending the Alarm

Once, when I was approaching a troop in the Huachucas, I got trapped in the bottom of a side canyon by coatis quickly headed in my direction. Normally, I would climb up one of the sides of the canyon, so I could watch them as they moved by. But in this instance, they were moving too fast, and the sides of the canyon were too open and steep for me to climb up fast enough to not be detected. So, I stood very still. The troop, with a bunch of 3-month-old youngsters, came barreling down the canyon, and several individuals passed right by me as I held my breath, not moving a muscle. One small youngster came within centimeters of my foot and then stopped and looked up at me. It fell over on its side, mouth opening and closing with an expression of abject terror. I heard no sound from the little one, but as I resisted the urge to pick it up and comfort it, an adult female rushed over, scooped it up with a foreleg, and ran off up the hill. At the same time, an alarm went out, and I heard several barks as the coatis scrambled and jumped into trees. The closest coati began alarm calling, and a minute or so later the other coatis descended from the trees and fled, all heading east.

When most of the troop had left, the calling coati followed the group, and within seconds the forest was quiet. I think the youngster that I frightened might have given out an ultrasonic call. It was the only instance I noted in which a nearby coati was opening and closing its mouth and I heard no sound. The maximum frequency we can hear declines with age, so it is possible that the calls were under 20 kHz (technically not ultrasonic then), but higher than I could hear, which was probably about 12 kHz at the time.

Because coatis in Arizona are usually quite shy, approaching too close typically resulted in coatis issuing alarm barks and making a rapid retreat from the area. The most often observed behaviors are those used by coatis to tell other coatis something dangerous is in the area and to make a retreat. Coatis indicate alarm using both visual and auditory displays. It is also possible that odor is involved, but that aspect remains unstudied.

Much like other coati signals, responses to sensed danger are graded – the intensity of the signal depends on the perceived amount of danger. Unless it is startled, a coatis' first response to something scary or unusual is to stop moving and start sniffing the air. If one animal stops to sniff, others often do the same, and the sudden quiet also allows coatis to hear if anything is amiss. Sometimes a coati will stand up on its hind legs to get a better view and perhaps increase its ability to detect scent. It may bob its head up and down, helping increase depth perception against a complex background of leaves and branches or perhaps giving its nose better exposure to odors or increasing its range of hearing.[1] If everything is ok, one of the elders (who was probably one of the first to stop moving) resumes foraging and the others soon follow her lead. The sound of her foraging and her contact calls assure the others that all is well.[2] If the source of alarm is serious but far away, the coatis will quietly retreat away from the danger. They stay close together, do not vocalize, and move quietly with their tails down; obviously doing everything they can to avoid detection. Yellow mongooses exhibited similar behavior.[3]

Visual Displays

When frightened or startled, coatis raise their tails upright and poof out the fur on their bodies and tails (piloerection), enhancing their apparent size and the visibility of the bands on the tail. A quick change in tail position, to upright or lowered, may serve to attract the attention of other coatis.[1] Occasionally, coatis will respond this way to small animals they haven't encountered before, like turtles or snakes.[1] Solo coatis are less likely to carry their tails vertically when they are foraging, indicating that the tails are probably used for communicating with other coatis when foraging in a group. Because solo coatis are much more vulnerable to

Female coatis using a perch to scan for danger. In this case, an adult male was approaching the troop. Guatemala, 2007.

predators, we might expect that they would avoid trying to draw attention to themselves, as an upright tail moving through the brush might do. This appears to be the case, although I am unaware of any studies that have quantified it. Coatis also use their tails as a visual alarm signal – lashing their tails back and forth if they are in a tree with something of concern below (as when I observed several coatis in a tree above a mountain lion). This serves to not only get the attention of nearby coatis but may also act as a predator-deterrent, letting the predator know it's been seen.[4,5] A predator that has lost the advantage of stealth is more likely to give up stalking and move on to other prey.[5]

Vocal Displays

If retreat is not an option, or the coatis are startled by an unknown danger, they take to the trees or cliffs. Usually a quick snort is issued, followed by the sound of claws on bark as the coatis scramble up the nearest trees. It is unknown if the vocalization is simply a startle response ("eek!") or a command to move quickly ("run!"). If the source of danger is unknown, they only climb high enough to look around, about 2-3 m (6-10 ft) or so. Here they cling to the side of a tree and hold still until the cause of the alarm is found. Once the threat is identified, one or more adults

155

closest to the threat start barking. Even baby coatis, newly emerged from the nest, will quickly climb up a tree if their mother issues an alarm call.

I was able to examine the structure of alarm barks from several video sources; the Ramsey video recorded in October 1994, and video made by my assistant, Jason, of two different females as they left their dens with their new litters. In addition, Mike Foster, a videographer from Bisbee, AZ, shared some video he shot of coatis at the San Pedro Riparian National Conservation Area. Alarm barks consisted of harsh broadband calls, repeated at a rate of about 3-6 per second, and sounded like a staccato cough or laugh, a raspy he-he-he-he-he-he-he. Although Kaufmann suggested that it was merely a rapid version of the contact grunt,[1] structurally it was quite different, and was made while rapidly opening and closing the mouth. The frequency ranged from 3 kHz to 14 kHz, rarely as low as 2 kHz, although they often included a separate but simultaneious grunt component under 0.1 kHz. The calling coati usually issued 7-14 barks, and then paused. They may repeat the series for minutes at a time. The bark itself was one of the least variable calls in structure, although it sometimes had other components added which might modify the message. Often a series of alarm barks was preceded by a chirp-click-grunt. Sometimes a snarl was added to the series of barks. The alarm barks of white-nosed and brown-nosed coatis were nearly identical.[6]

Chirp components added to barks may be a way to identify the signaler. Subadults would occasionally bark as I was trying to sneak up on troops, but they were usually ignored, indicating that either there may be a learning phase involved in using the proper alarm calls at the proper time, or caller identity or class (adults vs. subadults, for example) is important in deciding how to respond.[7,8] I was unable to examine the spectrograms of any subadult alarm calls to determine if they differed in structure from those of adults. Obviously, there is a lot of potential communication going on with this signal, but little is known about it. It is hard to study it in captive coatis because they seldom issue alarm barks.

A number of species give different types of alarm calls depending on the type of threat.[9,10] For example, some squirrels use different calls for aerial (e.g., hawk) versus terrestrial (e.g., fox) predators[11,12] and vervet monkeys have different calls for leopards, eagles, and pythons. An alarm call for leopard results in the monkeys running into the trees, one for eagle results in monkeys looking up, and one for python results in monkeys looking down.[13] These calls tend to be very loud. Coatis do not appear to have an alarm call like that, but rather a call that serves as predator-deterrent and lets the group know where the predator is. In many cases, the animal closest to me barked while perched part way up a tree and continued calling as the rest of the group retreated. In other cases, the caller descended the tree first and led the group away, usually with a series of chirp-grunts. Most of my observations were of coatis reacting to me or other humans. Human observers often commented that the coatis appeared to be "throwing insults" at them,[14] and there is some truth to that – the bark, like the tail-lashing described above, notifies the predator that it's been seen.[5,15] Among a wide variety of animals, communication with predators appeared to be the most important factor in the evolution of alarm calls.[5,15]

Coatis typically react to a human by running away on the ground. But they do not react that way to canid or felid predators, where they tend to climb higher in the trees and stay there. There may be a mixed response here, with coatis closest to the threat staying in the trees, and ones further away running off. Coatis, especially juveniles, will often run to the end of low branches and jump to the ground, avoiding a predator that might be trying to climb up the tree. How the message to run or climb is communicated is not known, but it is possible that there is another alarm call that has not been described yet.

In some cases, coatis respond to potential predators by climbing high in trees and running from branch to branch before descending and running away. One male (M39) would respond to people that way. He liked to rest in some tall sycamores right above a small picnic area at Fort Huachuca. When people would show up, he would quickly and quietly run through the treetops, descend quickly to the ground, and run away. But this behavior turned out to be detrimental when he panicked one too many times during a holiday weekend, when many people were using the picnic area. I found him dead a day later, curled up in the hollow at the base of one of the trees. When one of the veterinarians at Fort Huachuca examined him, he found that the coati had several broken bones and died of trauma consistent with a fall from a considerable height.

Some mammals, including yellow mongooses, and some species of ground squirrels, give calls that indicate not what the threat is, but how urgent it is.[3,16-18] They may do this by using a different call type for distant versus close predators, or change the call rate or amplitude as the threat gets closer.[11] Richardson's ground squirrels, for example, will use an ultrasonic call (48 kHz) when predators are spotted far away, but switch to an audible 8 kHz call when predators are close. Because ultrasonic calls don't travel very far, this allows the squirrels to notify just those animals close by (most likely their offspring) without giving themselves away to the predator.[18] The audible call, then, would alert the predator that it has been seen, and the location of the caller(s) would give an indication of the location of the predator.[19,20] No one has recorded coati alarm calls with equipment capable of recording ultrasounds. Given their use of ultrasounds in their contact calls (Chapter 10), it is possible that there are ultrasonic components to their alarm calls, too.

Among many animals, different types of alarm calls are innate, that is, youngsters are born capable of making the different calls, but they need to learn from adults when to use them.[7,19,21] Juvenile meerkats did not issue alarm calls very often, but when they did, it was often at non-dangerous animals in their environment.[22] Among meerkats, alarm calls contain information on the identity of the sender, although they usually responded the same way to any calling adult. They were less likely to respond to the alarm calls of juveniles.[23] I made similar observations of the coatis. Juveniles did not give out alarm barks, but they did give out distress and mobbing calls (chitters and loud chitters). They did not start using alarm barks until they were over a year old and were often ignored. Whether this was due to the identity of the caller, the message, or lack of development of the vocal apparatus is unknown.

Although barks were relatively simple in structure, a calling animal was able to potentially transmit a substantial amount of information. Based on my observations and examination of the spectrograms of a few barks, it appears that a member of the troop, hearing another coati barking, could determine the direction of the threat relative to their location, how serious the threat was (is it ok to go back to foraging or do we need to escape right now or do we need more information), what kind of threat it was (do we run or climb), and, possibly, who was doing the barking. The source of the alarm could also hear the bark, and thus was now aware that the coatis were aware of it. Coatis responded differently to different types of predators – they climbed to escape from large cats, they ran to escape from humans, and they dropped to the ground to escape from eagles. Coatis could potentially be encoding information about the type of predator (terrestrial or avian), perhaps by changing the call rate or frequency of the alarm calls, or by adding additional components, like chirps or snarls, but more research is needed to verify this.

Solo male coatis produced similar alarm barks to those made by females in troops. They also included contact calls as part of their barks, as well as the tail-waving display. Some of this was, no doubt, for the predator's benefit, but the inclusion of contact calls is interesting. Some males are acting as if they are part of a troop, even if there are no other coatis around.

Coatis differed in their responses to potential threats based on reproductive condition and location. Most troops were quite shy; Troop 1 in Ramsey Canyon was the most tolerant of people. They usually ignored people hiking by or near the visitor center. But away from the preserve, they were just as shy as all the other troops I was monitoring. As mentioned previously, females that were off by themselves while they had small babies in the nest would allow people to get quite close, possibly to keep an eye on them. However, when babies were back in the troop, the adult females were quite timid and protective of the youngsters. Males were more likely to become habituated to humans than females, and more likely to become problems by getting into people's yards, digging up their gardens and getting into trash receptacles. However, a couple of my radio-collared male coatis were extremely shy, and even with help of radio collars, I was seldom able to sneak close enough to see them.

Playbacks of barks at the Arizona-Sonora Desert Museum caused the coatis to move away from the source of the sound, where they quickly climbed up a large rock in their enclosure and quietly scanned the area. Higher quality recordings of coatis responding to different sources of alarms, as well as playbacks are needed before we can determine what kinds of information are encoded in their alarm calls.

Coatis also paid attention to sound of other animals' alarm calls. They have been observed being more vigilant after hearing alarm calls from birds and other animals.[1] I had heard from a couple of different hunters that when they were playing predator calls (recorded vocalizations of an injured animal designed to draw the attention and approach of a predator, usually a fox or coyote), sometimes coatis would approach them. Thinking this might be one way to get close enough to mark coatis, I obtained some commercial recordings of a wounded rabbit, snuck up close to a couple of different troops, and broadcast the recording while remaining hidden. In both cases, the troop members made panicked flights from the area, not even bothering to jump into trees to scan. I stopped the experiment after two sessions, not wanting to make the coatis even more afraid of me than they already were. When I spoke with one of the hunters again and got more details about seeing coatis coming into the predator call, it became apparent that only solo males were coming in. So, while there appears to be little differences in alarm calls between males and females, there may be a difference in response to the sound of a predator nearby. The adult females want to get their babies as far from a predator as possible, while solo males may be curious if there might be a morsel to steal. All of this is, of course, based on anecdote and conjecture, but I find the observations interesting, nonetheless.

Non-vocal Auditory Communication

Coatis produce additional sounds that may function in information transfer. When coatis are startled, say for example by a researcher trying to sneak up on them, they bound loudly away through the forest litter, then jump up into a tree to assess the situation. The sound of their bounding and the claws on the tree were very audible to me and let me know that there was no point in being sneaky anymore. I assume coatis were also tuned in to this noise. I think these sounds were an incidental part of their escape, and not intentional. On the other hand, coatis may pay little attention to things that sounds like leaves rustling or branches breaking. One time I was carefully sneaking up on Troop 3 just after they had regrouped after giving birth. As I bent down to crawl under the low branches of a pinyon tree, not 15 meters away from the group, my knees cracked loudly. I froze and held my breath for a few moments, expecting to hear a few alarm calls followed by the sounds of coatis crashing through the brush. But none of those sounds arose; instead, I could hear some soft chitters from the new youngsters. I continued creeping forward and was lucky enough to watch F28 roll over on her back and start nursing five kits. This was especially noteworthy, as I had seen her emerge from her birth den with only four kits, so she was nursing an additional kit besides her own (assuming four of the kits were hers). This has been reported in other populations of coatis before, but it was the first and only time I was able to get close enough to observe it in the Huachucas.

Silent Cues

Because a foraging group of coatis makes so much noise, both the noise of digging through the leaf litter and their constant vocalizations, sudden quiet is also noticeable. Although the foraging coatis often have their heads down with their noses buried in the leaf litter or stuck in holes, the contact calls and rustle of the leaf litter seems to serve as an "all is well" signal.[2,14] However, if the lead adult coati stopped vocalizing and moving, often everybody stopped (except some of the youngsters). This cessation of movement and vocalizing may last from a few seconds to several minutes, as the coatis stopped foraging and started looking around and sniffing the air. This allows the animals in the troop to hear if anything is trying to sneak up on them, as well as focusing their noses for predator detection.[24] It is possible that ultrasonic vocalizations are issued here, perhaps to coordinate behavior about when to stop and listen and when it is ok to go back to foraging, but we have no data on that yet.

Coatis would also appear to hide by going quiet. Often a troop of coatis would make a noisy retreat away from a threat, crashing through the brush and vocalizing, but other times, either when an animal was solo or it accidentally got left behind by the troop when they escaped, it would quietly hide from the observer and stay motionless.[25] I discovered this by radio-tracking, occasionally finding myself quite close to an animal hiding in the brush very quietly; only the signal from its radio collar giving it away. I also watched a couple of times as foraging coatis would freeze, when they heard people talking as they passed by on close, but out of sight, trails.

Notes and References

1. Kaufmann JH (1962) Ecology and social behavior of the coati, *Nasua narica*, on Barro Colorado Island, Panama. University of California Publications in Zoology 60:95–222

2. Townsend SW, Zöttl M, Manser MB (2011) All clear? Meerkats attend to contextual information in close calls to coordinate vigilance. Behavioral Ecology and Sociobiology 65:1927–1934

3. Le Roux A, Cherry MI, Manser MB (2009) The vocal repertoire in a solitary foraging carnivore, *Cynictis penicillata*, may reflect facultative sociality. Naturwissenschaften 96:575–584

4. Hass CC, Valenzuela D (2002) Anti-predator benefits of group living in white-nosed coatis (*Nasua narica*). Behavioral Ecology and Sociobiology 51:570–578

5. Hasson O (1991) Pursuit-deterrent signals: communication between prey and predator. Trends in Ecology and Evolution 6:325–329

6. Gasco A, Ferro HF, Monticelli PF (2018) The communicative life of a social carnivore: acoustic repertoire of the ring-tailed coati (*Nasua nasua*). Bioacoustics 28:459–487

7. Hollen LI, Manser MB (2006) Ontogeny of alarm call responses in meerkats, *Suricatta suricatta*: the roles of age, sex and nearby conspecifics. Animal Behaviour 72:1345–1353

8. Ramakrishnan U, G. Coss R (2000) Age differences in the responses to adult and juvenile alarm calls by bonnet macaques (*Macaca radiata*). Ethology 106:131–144

9. Manser MB, Seyfarth RM, Cheney DL (2002) Suricate alarms signal predator class and urgency. Trends in Cognitive Sciences 6:55–57

10. Sieving KE, Hetrick SA, Avery ML (2010) The versatility of graded acoustic measures in classification of predation threats by the Tufted Titmouse *Baeolophus bicolor*: exploring a mixed framework for threat communication. Oikos 119:264–276

11. Blumstein DT (2007) The evolution of alarm communication in rodents: structure, function, and the puzzle of apparently altruistic calling. *In*: Wolff JO, Sherman PW (eds) Rodent societies: an ecological and evolutionary perspective. University of Chicago Press, Chicago, IL, pp 317–341

12. Kiriazis J, Slobodchikoff CN (2006) Perceptual specificity in the alarm calls of Gunnison's prairie dogs. Behavioural Processes 73:29–35

13. Seyfarth RM, Cheney DL, Marler P (1980) Monkey responses to three different alarm calls: evidence of predator classification and semantic communication. Science 210:801–803

14. Gilbert B (1973) Chulo. A year among the coatimundis. Alfred A. Knopf, New York, NY

15. Shelley EL, Blumstein DT (2005) The evolution of vocal alarm communication in rodents. Behavioral Ecology 16:169–177

16. Blumstein DT (1999) Alarm calling in three species of marmots. Behaviour 136:731–757

17. Furrer RD, Manser MB (2009) The evolution of urgency-based and functionally referential alarm calls in ground-dwelling species. The American Naturalist 173:400–410

18. Wilson DR, Hare JF (2006) The adaptive utility of Richardson's ground squirrel (*Spermophilus richardsonii*) short-range ultrasonic alarm calls. Canadian Journal of Zoology 84:1322–1330

19. Slobodchikoff C (2012) Chasing Doctor Dolittle: learning the language of animals. St. Martin's Press, New York, NY

20. Personal observation.

21. Gouzoules H, Gouzoules S (1989) Design features and developmental modification of pigtail macaque, *Macaca nemestrina*, agonistic screams. Animal Behaviour 37:383–401

22. Hollen LI, Clutton-Brock TH, Manser MB (2008) Ontogenetic changes in alarm-call production and usage in meerkats (*Suricatta suricatta*): adaptations or constraints. Behavioral Ecology and Sociobiology 62:821–829

23. Schibler F, Manser MB (2007) The irrelevance of individual discrimination in meerkat alarm calls. Animal Behaviour 74:1259–1268

24. Di Blanco Y, Hirsch BT (2006) Determinants of vigilance behavior in the ring-tailed coati (*Nasua nasua*): the importance of within-group spatial position. Behavioral Ecology and Sociobiology 61:173–182

25. Also observed by Kaufmann (1962).

13

What to Eat

oatis live in a three-dimensional world, living comfortably on the
ground, but also scaling trees and cliffs with ease. They use long
claws on their front feet for digging through the leaf litter, turning
over rocks, and digging up delicacies like tarantulas. They readily ascend
trees in search of fruit. They poke their long, flexible nose into holes,
cracks, and crevices in search of edibles. Most of their day, from a little
after sunrise to sunset, is spent foraging for food.

In addition to following coatis' telemetry collars, I also spent a lot of
time looking for coati sign. I studied their trails, scrutinized their tracks,
observed their scratches on trees and agaves, and sniffed a lot of scats[1] to
learn to distinguish between coati and gray fox scat. I examined how they
tilled through the leaf litter and through the snow, how they turned over
rocks and cow pies, and how they left unusual "u"-shaped marks in the
dust with their noses. I investigated the nests they made in cottonwood
and ash trees and climbed up cliffs to examine the cracks and crevices
where they spent the night. Telemetry gave me information on where
they were and how they were moving; tracking let me know what they
were doing in those places. And all this information contributed to my
understanding of how they moved, where they slept, and what they ate.

Coati tracks look like small bear tracks. I do not think this is a coinci-
dence as coatis inhabit similar environments and eat similar foods. They
spend their days turning over rocks, digging up bugs and climbing trees to

Coati tracks in soft dust, left, right, "nose circles", left by coatis sniffing the dirt. Direction of travel is toward the bottom of the photograph.

get to fruit. They amble along like bears do, and their feet are shaped similarly. Over the years I began to think of coatis as little bears that live in gangs. In German, the word for coati is "Nasenbären," or nose-bear,[2] which is a pretty good description.

What Is on the Menu?

Coatis are considered omnivorous, which means they eat basically anything. This might be overstating things. In captivity, they will eat most things given to them, with a preference for sweets, but in the wild their diet is predominantly fruit and bugs. Some populations have access to sea turtle eggs or other high-protein goodies and will take advantage of them, but almost all studies have shown that coatis focus on fruit and bugs, with the composition of their diets depending upon where they live. Studies have found that most food is found using their sense of smell.[3,4] Small vertebrates, including snakes, lizards, frogs, turtles, mice, and birds are also consumed, but the coatis' bread-and-butter is a diet of fruit and bugs.

Figuring out what coatis eat is not an easy task, even for researchers working on habituated coatis. It is not difficult if coatis are up in a tree munching on fruit, but when their entire head is buried in the leaf litter, it is hard to tell if they are going after a bug or an old rotten piece of fruit.

Assessing the coati's diet, then, takes not only field observations, but scat and stomach analyses in the lab, too. Unlike many animals that create latrines as a form of message post, coatis defecate as they are foraging and moving from place to place, making scats difficult to find in the leaf litter.

Left, In deep leaf litter, it can be difficult to tell what coatis are eating. Right, eating an earthworm. Photo by Ben Hirsch.

Coatis will also defecate after resting, either after rising in the morning, or after their mid-day siesta. I collected scats opportunistically, by carefully watching animals as they foraged, as well as collecting any deposits left behind in traps. Coati scats look a lot like gray fox scats, but they do not have the musky odor that gray fox scats do. Gray foxes often leave scats on top of rocks and logs, so I avoided picking up scats in those locations. Collected scats were frozen and sterilized before I picked them apart and examined the contents under a dissecting scope and a microscope. I made a reference collection of seeds, insects, bird feathers, reptile scales, and bones to compare with the contents of coati scats, and used published guides to mammal hair, bones, and insects.[5-7]

I collected 144 scats from live animals, and stomachs of three animals that died due to various causes. The contents of the stomachs and scats were combined for analysis, and here I just refer to them all as "scats." I present the results as the proportion of the total number of scats that contain a particular food item or category. Most scats had more than one type of food item. Overall, 87% of scats contained fruits, 78% contained invertebrates (snails, earthworms, and insects), and only 3% contained vertebrates (rabbits, mice, small birds, and lizards). Types of foods in scats varied seasonally, although perhaps not as much as you might expect. In all seasons except spring (April to June), coati scats contained mostly fruits, closely followed by invertebrates. During spring, the ratios were reversed, with slightly more invertebrates consumed than fruits. This makes sense, as fewer fruits were available to the coatis during that season.

Nine different fruits were found in coati scats. Throughout the year, fruits of alligator junipers were the most common items consumed, and the percentage of scats containing juniper berries ranged from a low of 43% during the 2nd quarter (dry season – April to June) to a maximum of 95% during the 1st quarter (winter – January to March). Some fruits were only available seasonally, with barberry consumed in late spring; buckthorn, prickly pear, canyon grape, and chokecherry consumed in mid to late summer; and madrone consumed during late fall and early winter. Other fruits either had longer fruiting seasons, such as manzanita, or had hard seed coats like acorns, and were available over several seasons.

The amount of juniper berries consumed by coatis is particularly interesting considering their medicinal properties. Berries and leaves of various species of junipers have been used medicinally by humans for millennia, for treating everything from arthritis and bruises to preventing malaria. The secondary compounds in the berries, which are also used to flavor gin, stimulate the kidneys and can be quite powerful diuretics.[8] So how do coatis (and black bears, gray foxes, wild turkeys, and all the other animals that eat a lot of alligator juniper berries) manage to eat so many without destroying their kidneys? No one knows for sure, but alligator juniper berries appear to be lower in kidney-damaging volatile oils than other juniper berries.[9] I tasted several kinds of local juniper berries, and mature alligator juniper berries were much sweeter and milder tasting than the others. Alligator juniper berries have a hard shell when they mature, and germination is enhanced by passing through the digestive tracts of animals, especially birds,[9] but I suspect coatis as well. Alligator juniper seeds survive a coati's digestive system mostly intact, although the outer husks were usually cracked into pieces (which is how I was able to identify them in scat). Thus, coatis and other animals are probable seed dispersers for alligator juniper.

On several occasions during the summer, I saw female coatis gorge on birchleaf buckthorn berries, then run over to a juniper tree and start rapidly eating green (immature) juniper berries. I found the buckthorn berries to be a bit tart, and other species of buckthorn are known for their laxative effects. I do not know if coatis were self-medicating with the berries of either of these species, but it was a curious observation.

Coatis are one of many tropical frugivores (fruit-eaters) that are important seed dispersers.[10-13] Many tropical trees produce fleshy fruits that are attractive to birds and mammals, which consume the fruits and carry the seeds away from the parent plant. Studies of fruits eaten by coatis have found that many are capable of germinating following digestion,

Male white-nosed coati climbing a large alligator juniper to get the berries.

although no one has identified any plants that depend fully on coatis for seed dispersal.[11,12] However, their role as seed dispersers may be more important in secondary forest and forest edges, as they may be more likely to use those habitats than other frugivores.[10] Coatis may be important in reestablishing forests after logging or fire.

The most common invertebrates in coati diets were beetles and beetle larvae (Coleoptera, about 75% of scats). Coleopterans made up at least 50% of the diet in all seasons except during the 2nd quarter. During that time, they increased their consumption of caterpillars of moths and butterflies (Lepidoptera), desert centipedes (Chilopoda), ants and termites (Hymenoptera), and grasshoppers, crickets, and cicadas (Orthoptera). After the rains began in July, they increased their consumption of land snails (gastropods).

Coatis ate a variety of other invertebrate taxa, including earthworms, scorpions, spiders and tarantulas, dragonfly larvae (odonates), true bugs, flies, termites, and ants. But none of these groups were found in more than 10% of scats in any season. Coatis rapidly roll tarantulas under their front feet to remove the hairs before consuming them.[14] Coatis in my study rarely consumed vertebrates, with the remains of a rabbit, mouse, and bird found in one scat each, and lizard remains showing up in two scats. Lizards only appeared in scats during the winter, when they may have been easier to catch due to lower temperatures. Kaufmann speculated that lizards might be very important in the diets of coatis in Arizona,[15] but so far, the data haven't supported his conjecture. However, coatis are opportunistic, and it is likely that the proportions of different foods in the diet vary with what is available in different areas. For example, coatis living along the major stream habitats in southeastern Arizona (e.g., San Pedro River, Aravaipa Creek) have little access to alligator juniper but hackberry trees are present, and the coatis there consume a lot of hackberry fruits. But hackberries are seasonal, and in those areas the coatis appear to spend more of their time feeding in the leaf litter, most likely on beetles and crickets, but more study is needed on this.

In addition to the variety of invertebrates and fruits found in scats, I observed coatis eating some things that did not appear in scats. During the summer, coatis often climbed up large flowering agaves (century plants) and lapped up the nectar. They may have also obtained some insects attracted to the flowers,[16] but I think what they were seeking was the sweet and plentiful nectar produced by the plants. The two most common species of agaves in the Madrean woodlands of the Huachucas are Parry's agave and Palmer's agave. Both are suspected to be pollinated by bats but appear adapted to multiple pollinators.[17] I think terrestrial mammals may be important pollinators for the Parry's agaves, which are very robust and tolerate being climbed on and having their nectaries raided by coatis with little damage to the flowers. I have watched female coatis climbing Parry's agaves, carefully probing their long noses into the inflorescences, moving from flower to flower before descending to find another plant. They acquire enough pollen during this nectar feeding to turn their face and throat bright yellow, and they are probably pollinating the flowers as they move from plant to plant. Palmer's agaves, on the other hand, are more delicate and suffer substantial flower damage and loss, including entire branches broken, when coatis climb them.

Flowering agaves were so attractive to coatis that a crude survey for the presence of coatis could be done simply by examining the stalks of agaves, post-blooming, for scratches made by their hind feet as they descended the plants. Although agaves were also climbed by squirrels and ringtails, they leave smaller scratches. Raccoons would leave similar-sized scratches, but there were few raccoons in the areas with Parry's agaves.

Left, Adult female coati descending after feeding on nectar in a flowering Parry's agave, July 2000. Right, scratches left on agave stalk by coatis.

However, because of that possibility, a lack of scratches on agaves likely meant there were no coatis in the area, while the presence of scratches merited further investigation.

Similar observations were made of coatis pollinating balsa trees in Costa Rica.[18] And brown-nosed coatis were one of many species of birds and bats pollinating *Mabea fistulifera* in Brazil.[19] Not very many observations of pollination by coatis have been made, and in all cases, they appear to be opportunistic; no plants are known to depend exclusively on coatis for pollination.

Other foods that I observed being eaten that did not appear in scats included cottonwood and chokecherry flowers, where they appeared to be consuming whole flowers, and earthworms. They dug the earthworms out of the soil and tugged them out of their burrows much like a robin does and looked like they were slurping strands of spaghetti.

Coatis are fond of carrion, and it is possible that the rabbit and bird remains I found in coati scats were from carrion – that is, they found the animals already dead, as there were no bones in the scat, only hair or feathers. With animals of this size, I would expect them to consume all or most of the carcass, including bones. Feathers and bones were found in scats of coatis studied in Brazil.[10] I have also observed coatis feeding on the remains of a white-tailed deer (apparently killed by a mountain lion). In this case, I followed the telemetry signal of an adult male, and when I caught up to him, he was feeding on a deer carcass at the Ramsey Canyon Nature Conservancy Preserve. He left when he detected me. I asked my assistant, Jason, to keep an eye on the site for the next few days, and he reported that the local troop of coatis (Troop 1) discovered the carcass within 48 hours and dismembered it and carried pieces off into the brush. When I returned to the area a couple of weeks later, there was nothing left at the site except for a few hairs.

I also radio-tracked an adult female coati to a black bear carcass one summer. There was little left of it; it had been already picked over and was quite rotten. In this case, at least while I was watching, she did not feed on the carcass, but overturned parts and fed on the dermestid beetles that were breaking the carcass down. The attraction of coatis to carrion (observed by other researchers as well) made me wonder if some of the coatis that were killed by mountain lions could have occurred at sites where lions had killed other prey. Lions are known to attack and kill other lions and bobcats that show up at their kill sites,[20-22] so it seems possible that coatis might be vulnerable in this situation, too. However, I only found one radiocollared coati that was killed by a lion near a deer carcass, which was too completely consumed to figure out what killed it.

I just missed seeing an actual case of coati predation on a vertebrate. During the summer of 1998, I was tracking F149 as she was out foraging away from the nest holding her young kits. As I approached her signal, I heard a tremendous ruckus, with a lot of squawking, and saw a pair of Whiskered Screech-Owls dive bombing something. As I looked closer, I saw F149 at the base of a tree devouring a baby owl (owlet). Nearby, another owlet, newly fledged, repeatedly squawked from a low branch. The parents screeched as they repeatedly dive-bombed F149. But it was too late, the chick was dead, and F149 responded to the dive-bombing parents with nose-up displays as she consumed the entire owlet. When she was finished, she left and resumed foraging. The adult owls coaxed their remaining chick to follow them and left the area. Although tragic for the owls, F149 was nursing young of her own at the time, and the high-protein meal was no doubt of benefit. When I checked the area where F149 had consumed the owlet, I found no remains, not even a single feather.

Baby coatis are entirely dependent on their mother's milk until they leave the natal den at about 5-6 weeks of age. The coati mama does not bring any food back to the den. As soon as they leave the den, they start foraging alongside the adults, although they still nurse for several more months from their mothers and, on rare occasion, from other lactating females in the troop. Females may occasionally give up a morsel to their young kits, but milk is the only food provided to the kits; the rest they must dig up on their own. Kits are often observed sniffing and licking their mother's mouths; in this way, they may learn what foods are most palatable.[23,24] The kits seem capable of eating the adult diet shortly after leaving the den, and I've heard baby coatis crunching ripe juniper berries right alongside the adults.

The results of my study in the Huachucas are comparable to food habits studies done on coatis elsewhere. In Panama, Guatemala, Mexico, Costa Rica, and Argentina, coatis ate mostly fruits and invertebrates, with less than 2% of the diet being vertebrates.[12,23,25,26] However, on the west coast of Mexico, up to 15% of scats contained mammals, birds, and reptiles, mostly sea turtle eggs.[27] Coatis are also important predators of sea turtle eggs in Costa Rica.[28] The diets of brown-nosed coatis were also predominantly fruit and bugs,[29-31] although one study in Brazil found almost 10% of scats contained vertebrates, mostly birds.[10] They also obtain some of their food by consuming the invertebrates that collect within bromeliads.[32] Of course, the foods consumed depend on local availability, and more tropical-living coatis get to enjoy papaya, mangos, palm nuts, figs, and land crabs; foods unavailable to their Arizona relatives. Some studies in the tropics have found that they eat more than 50 different kinds of fruits.[10] However, they have more competitors, and are often chased out of trees by capuchin monkeys.[14]

One study of the diets of mountain coatis found a higher percentage of invertebrates in the diet (100% of scats). Frogs (55% of scats), fruits (37%), and vegetable matter (41%) also comprised the diet; reflecting the differences in food availability high in the Andes.[33] Compared to other carnivorans, coatis do not appear to depend much on vertebrate prey. My impression from handling coatis and the skulls of dead coatis was that their jaws were quite delicate, much more so than those of raccoons. This has been backed up by studies of bite forces of carnivorans (how much pressure they can apply with their teeth); coatis have among the lowest bite force for their size, and less than half that of raccoons.[34] Coatis are not built for killing animals much bigger than bugs.

Food Availability

Curious as to how the diet reflected what was available in the local environment, my assistant, Jason, and I also studied the seasonal availability of fruits and invertebrates in several major habitat types on Fort Huachuca. Every two weeks from March 1996 to February 1997 we went out and sampled four different transects, one each in riparian forest, chaparral, encinal, and oak-pine woodland.[35] I recorded the plant phenology (dormant, flowering, fruiting, etc.), and Jason carefully dug through the leaf litter counting bugs. Although we were unfortunate enough to conduct the study during an extreme drought, I think the results reflect the differences among seasons and habitats for coati food availability. In general, riparian habitats had the greatest availability of fruits and invertebrates during all seasons. Among habitats, counts of invertebrates were higher on transects with higher soil moisture. And, as might be expected, the availability of invertebrates in all habitats increased during the monsoon.[36] The fact that Jason was able to count the bugs on a transect is also an indication of the low resource availability at the northern end of their range. If we had tried this method in the tropics, Jason would still be counting. The greatest numbers and varieties of fruits were also available during the monsoon, but some invertebrates and fruits were available during all seasons, and this fact has major implications as to where coatis can live.

Surviving Food-limited Seasons

Coatis appear to lack the ability to store a lot of fat to use as an energy source during the winter, like their raccoon cousins do.[37] It takes a specially-adapted physiology to tolerate extended dormancy (torpor), including the ability to slow down metabolic pathways that will start digesting muscle for energy and generate toxic levels of waste by-products. Raccoons in the colder parts of their range will take extended naps for days to months at a time during cold or inclement weather, although they may come out and forage during warmer days.[38-40] They may lose 50% of their body mass during the winter. In Arizona, perhaps the coldest habitats that coatis live in, they did not fatten and fast during the winter, rather males lost weight during the mating season (March-April) and females lost weight during lactation (July-August).[36] Coatis appear to need to eat every day, particularly if it's cold outside. So, they need to live in areas where

Coatis foraging in the snow in upper Garden Canyon, 1998.

fruit and bugs are available year-round. This may limit their northward expansion, as snow and especially ice may limit their access to the leaf litter. Although coatis can dig through several inches of snow, I think a thick coat of ice would create serious problems for them.

Much of coati range, from Mexico to South America, is classified as Tropical Dry Forest. This habitat type is characterized by extreme seasonality, including dry seasons of 3-8 months long.[41] These long dry seasons reduce the abundance and distribution of water and soil invertebrates.[42–45] In some areas, fruits are more abundant during the dry season.[43] In these highly seasonal habitats, coatis change their foraging and ranging behavior to compensate for the change in availability of resources.[43] But even when food or water are limited, coatis remain active year-round.

Several researchers have seen feeding associations between coatis and other animals. In Guatemala, White Hawks were observed following coatis as they foraged in the leaf litter, perhaps to pick off any lizards (their usual prey) that the coatis kicked up.[46] In Brazil, a variety of birds, including hawks, trogons, ant tanagers, and wood creepers[47] as well as collared peccary[48] were observed associating with foraging coatis. In the Huachucas, I saw wild turkeys in association with foraging coatis on a several occasions.

Notes and References

1. It is possible to inhale diseases from sniffing scats of some carnivores, so don't try this at home.

2. Michael Sutor, personal communication.

3. Gompper ME (1995) Nasua narica. Mammalian Species 487:1–10

4. Hirsch B (2010) Trade-off between travel speed and olfactory food detection in ring-tailed coatis (*Nasua nasua*). Ethology 116:671–679

5. Borror DJ, White RE (1970) A field guide to the insects. Houghton Mifflin Co., New York, NY

6. Moore TD, Spence LE, Dugnolle WG (1974) Identification of the dorsal guard hairs of some mammals of Wyoming. Wyoming Game and Fish Department, Lander

7. Olsen SJ (1968) Fish, amphibian and reptile remains from archaeological sites. Part 1: southeastern and southwestern United States. Papers of the Peobody Museum of Archeology and Ethnology 56:1–103

8. Kowalchik C, Hylton WH (1998) Rodale's illustrated encyclopedia of herbs. Rodale Books, Emmaus, PA

9. Tirmenstein D (1999) *Juniperus deppeana*. US Department of Agriculture, Forest Service, Rocky Mountain Research Station, Fire Sciences Laboratory

10. Alves-Costa CP, da Fonseca GAB, Christofaro C (2004) Variation in the diet of the brown-nosed coati (*Nasua nasua*) in southeastern Brazil. Journal of Mammalogy 85:478–482

11. Alves-Costa CP, Eterovick PC (2007) Seed dispersal services by coatis (*Nasua nasua*, Procyonidae) and their redundancy with other frugivores in southeastern Brazil. Acta Oecologica 32:77–92

12. Sáenz JM (1994) Ecología del pizote (*Nasua narica*) y su papel como dispersador de semillas en el bosque seco tropical, Costa Rica. Master's Thesis, Universidad Nacional, Heredia, Costa Rica

13. Wright SJ, Zeballos H, Domínguez I, Gallardo MM, Moreno MC, Ibáñez R (2000) Poachers alter mammal abundance, seed dispersal, and seed predation in a Neotropical forest. Conservation Biology 14:227–239

14. Kaufmann JH (1962) Ecology and social behavior of the coati, *Nasua narica*, on Barro Colorado Island, Panama. University of California Publications in Zoology 60:95–222

15. Kaufmann JH, Lanning DV, Poole S (1976) Current status and distribution of the coati in the United States. Journal of Mammalogy 57:621–637

16. McColgin ME, Brown EJ, Bickford SM, Eilers AL, Koprowski JL (2003) Use of century plants (*Agave palmeri*) by coatis (*Nasua narica*). The Southwestern Naturalist 48:722–725

17. Slauson LA (2000) Pollination biology of two chiropterophilus agaves in Arizona. American Journal of Botany 87:825–836

18. Mora JM, Mendez VV, Gomez LD (1999) White-nosed coati Nasua narica (Carnivora: Procyonidae) as a potential pollinator of *Ochroma pyramidale* (Bombaceae). Revista de Biologia Tropicale 47:719–721

19. Olmos F, Boulhosa RLP (2000) A meeting of opportunists: birds and other visitors to *Mabea fistulifera* (Euphorbiaceae) inflorescences. Ararajuba 8:93–98

20. Koehler GM, Hornocker MG (1991) Seasonal resource use among mountain lions, bobcats, and coyotes. Journal of Mammalogy 72:391–396

21. Hass CC (2009) Competition and coexistence in sympatric bobcats and pumas. Journal of Zoology 278:174–180

22. Hansen K (2006) Bobcat: master of survival. Oxford University Press, Oxford, UK

23. Russell JK (1982) Timing of reproduction by coatis (*Nasua narica*) in relation to fluctuations in food resources. *In:* Leigh EG Jr (ed) The ecology of a tropical forest. Seasonal rhythms and long-term changes. Smithsonian Institute Press, Washington, D.C., pp 413–431

24. Personal observation.

25. Binczik GA (2006) Reproductive biology of a tropical procyonid, the white-nosed coati. Ph.D. Dissertation, University of Florida, Gainesville

26. Delibes M, Hernandez L, Hiraldo F (1989) Comparative food habits of three carnivores in Western Sierra Madre, Mexico. Zeitschrift Fur Saugetierkunde - International Journal of Mammalian Biology 54:107–110

27. Valenzuela D (1998) Natural history of the white-nosed coati, *Nasua narica*, in the tropical dry forests of western Mexico. Revista Mexicana de Mastozoologica 3:26–44

28. Fowler LE (1979) Hatching success and nest predation in the green sea turtle, *Chelonia mydas*, at Tortuguero, Costa Rica. Ecology 60:946–955

29. Hirsch BT (2009) Seasonal variation in the diet of ring-tailed coatis (*Nasua nasua,*) in Iguazu, Argentina. Journal of Mammalogy 90:136–143

30. Aguiar LM, Moro-Rios RF, Silvestre T, Silva-Pereira JE, Bilski DR, Passos FC, Sekiama ML, Rocha VJ (2011) Diet of brown-nosed coatis and crab-eating raccoons from a mosaic landscape with exotic plantations in southern Brazil. Studies on Neotropical Fauna and Environment 46:153–161

31. Bianchi R de C, Campos RC, Xavier-Filho NL, Olifiers N, Gompper ME, Mourão G (2014) Intraspecific, interspecific, and seasonal differences in the diet of three mid-sized carnivores in a large neotropical wetland. Acta Theriol 59:13–23

32. Beisegel B de M (2001) Notes on the coati, *Nasua nasua* (Carnivora: Procyonidae) in an Atlantic forest area. Brazilian Journal of Biology 61:698–692

33. Rodriguez-Bolanos A, Cadena A, Sanchez P (2000) Trophic characteristics in social groups of the mountain coati, *Nasuella olivacea* (Carnivora: Procyonidae). Small Carnivore Conservation 23:1–6

34. Christiansen P, Adolfssen JS (2005) Bite forces, canine strength and skull allometry in carnivores (Mammalia, Carnivora). Journal of Zoology 266:133–151

35. Each transect was 100 m long, divided into four quadrats of 25 m. One 25-m quadrat from each transect was used for both fruit phenology and invertebrate abundance. To assess fruit phenology, all shrubs and trees within 5 m of the center line of the quadrat were examined and classified

into one or more phenological stages: non-reproductive, flower buds, flowers, immature fruit, ripe fruit, fruit-on-ground, and last-year's fruit. To assess invertebrate abundance, all invertebrates > 2 mm in length were counted within the top 10 cm of leaf litter along a 10 m long, 60 cm wide strip. Ants moving within the litter were counted; ant nests were not. Invertebrates were identified to order. All transects were sampled biweekly.

36. Hass CC (1997) Ecology of white-nosed coatis in the Huachuca Mountains, Arizona, a preliminary study. Final report submitted to the Arizona Game & Fish Department, Phoenix, AZ. 1–52

37. Mugaas JN, Seidensticker J, Mahlke-Johnson KP (1993) Metabolic adaptation to climate and distribution of the raccoon *Procyon lotor* and other Procyonidae. Smithsonian Contributions to Zoology 542:1–34

38. Mugaas JN, Seidensticker J (1993) Geographic variation of lean body mass and a model of its effect on the capacity of the raccoon to fatten and fast. Bulletin of the Florida Museum of Natural History, Biological Sciences 36:85–107

39. Mech LD, Barnes DM, Tester JR (1968) Seasonal weight changes, mortality, and population structure of raccoons in Minnesota. Journal of Mammalogy 49:63–73

40. Zeveloff SI (2002) Raccoons: a natural history. Smithsonian Books, Washington, D.C.

41. Murphy PG, Lugo AE (1986) Ecology of tropical dry forest. Annual Review of Ecology and Systematics 17:67–88

42. Levings SC, Windsor DM (1982) Seasonal and annual variation in litter arthropod populations. *In*: Leigh EG Jr (ed) The ecology of a tropical forest. Seasonal rhythms and long-term changes. Smithsonian Institution Press, Washington, D.C., pp 355–387

43. Valenzuela D, MacDonald DW (2002) Home-range use by white-nosed coati (*Nasua narica*): limited water and a test of the resource dispersion hypothesis. Journal of Zoology 258:247–256

44. Buskirk RE, Buskirk WH (1976) Changes in arthropod abundance in a highland Costa Rican forest. American Midland Naturalist 95:288–298

45. Janzen DH, Schoener TW (1967) Differences in insect abundance and diversity between wetter and drier sites during a tropical dry season. Ecology 49:96–110

46. Booth-Binczik SD, Binczik GA, Labisky RF (2004) A possible foraging association between White Hawks and White-nosed Coatis. Wilson Bulletin 116:101–103

47. Beisegel B de M (2007) Foraging association between coatis (*Nasua nasua*) and birds of the Atlantic Forest, Brazil. Biotropica 39:283–285

48. Desbiez ALJ, Rocha FL, Keuroghlian A (2010) Interspecific association between an ungulate and a carnivore or a primate. acta ethologica 13:137–139

14

Where to Live

Routine radio-tracking of the coatis provided a lot of information about where coatis live and how they move from place to place. When I began my study, it was thought that coatis in Arizona were "semi-nomadic," and that they had no fixed home ranges but rather moved extensively throughout the mountains, possibly migrating back into Mexico periodically. What we found was that coatis were not nomadic at all, but rather had vast and stable home ranges.

Home Ranges

The terms "home range" and "territory" are often used interchangeably, but they have different meanings to animal biologists. A home range is the area used by an animal in the usual course of its day-to-day activities. It includes seasonal movements but not dispersal or occasional forays to distant areas. A territory, on the other hand, is a home range, or part of a home range, that is actively defended against other members of the same species.[1] To actively defend an area requires more energy and perhaps different communication signals. Animals may have overlapping home ranges that are not actively defended but may adjust their behavior and activity to avoid conflict with neighbors. So, the form of space use, defended or not, is intimately tied with an animal's communication system as well as its need for resources, such as food, water, and shelter.

People also have home ranges. Imagine if you were to plot your location on a map every hour for a week. Notice how the dots on the map pile up for locations in which you spend a lot of time, say the bedroom, and your chair at the office. Notice that dots also include your travel routes between home and work, and occasional forays to the grocery store. Now imagine if we extend this mapping project out for a year. The area encompassed by the dots would probably get a little larger, as it would include more activities – visits to friends' houses, maybe travel for work or to visit relatives in another state, maybe a vacation on the other side of the country. But even though your home range map gets larger, most dots are still concentrated between your home and work. When we talk about a home range, we can divide it up by the frequency of dots in space. The total range encompasses all the dots for a particular period. The home range includes 90-95% of the dots (depending on the model used to calculate it), excluding the most unusual forays, like that once-in-a-lifetime photo safari in Africa. The core area includes only those areas used most, focusing on the critical use areas.

Home ranges may change over time, as you move out of your parent's house, or get a job in a different part of the country. If you spend summers up north and winters in the south (or vice-versa in the southern hemisphere), your home range would look quite different if it was calculated for just one season versus calculated for a whole year. My questions about space use concerned how large coati home ranges were (especially compared with those in the tropics), how they related to other troops (how much they overlapped, and did they show any form of territoriality), and how home range boundaries changed from season to season and from year to year.

Through repeated use, animals develop a mental, or cognitive, map of their environment.[2] This map is not only based on visual landmarks, but also incorporates smells, and possibly sounds and textures. Although poorly researched in wild animals, the cognitive map theory fits with what we know of how animals move through their environment.[3,4] The development and continual updating of a cognitive map allows an animal to quickly identify changes to their environment, for example, that unfamiliar object in the distance just might be a predator. One example for humans might be the layout of your house. Can you close your eyes and picture the route to the closest bathroom? That is an example of a cognitive map. A coatis' cognitive map would include important areas like water sources, food sources, denning areas, and perhaps areas where they are more likely to encounter predators (the "landscape of fear").[5,6] It probably also includes the locations of neighboring troops. Because resources change seasonally and year-to-year, their cognitive maps may also include a temporal component, such as information on where water can be found during the dry season, and when and where to find ripe chokecherries.

Radio-tracking coatis typically consisted of wandering the canyons and ridges using a telemetry receiver to scan the frequencies of coatis wearing radio collars. If we detected a signal, we tried to find the collared animals in one of two ways, which were not mutually exclusive. First, the signal usually gave us general information about direction and distance. From there, we would usually try to get close to the target animals. Using a procedure called triangulation, we could estimate a coati's location from one spot, then quickly (in less than 10 minutes) move to a new spot and estimate location again. If we drew a line between the spots where we stood and the coatis' signals, the intersection of those two lines was the estimated location for our dot on the map. We tested the accuracy of this method using test collars. One of us would take a radio collar out into the field (so we had a known location), and the other would estimate the location. I also ran this test for most mortalities – triangulating an estimated location before moving in on the mortality signal. Triangulating telemetry signals usually put our estimated locations within 41 m (130 ft) of the actual location; less at closer distances, more at further distances.[7] We combined triangulation with changes in signal strength to improve accuracy – for example, if triangulating told us the troop was somewhere on a slope, but the signal quickly disappeared, we could assume that the animals had crossed over the ridge and moved into a new canyon.

Triangulation allowed us to estimate locations on the animals without disturbing them. Current technology using GPS collars is more accurate, requires less field time, and may disturb the animals less. However, small enough GPS collars were not available during my study. And even GPS technology would give us only limited information about what the coatis were doing and who was in the group (unless every animal was collared). So, we used the radio signals to approach the groups, doing our best to carefully sneak within view while trying to disturb them as little as possible. With the troop's multiple sets of ears, eyes, and noses all working to detect predators, sneaking up on the coatis close enough to see them required good stalking skills. We had to pay attention not only to clues that might indicate a coati was nearby – rustling in the leaves, coati vocalizations, squawking of jays, – but also avoid revealing our presence. Coatis were not the only animals that rustled through the brush or attracted the attention of jays. I snuck up on quite a few wild turkeys, and once found myself almost nose-to-nose with a small bear cub. That required a quick retreat before the cub's mama discovered me. In addition to moving quietly, we had to pay attention to wind direction, and we had to be conscientious of our odors by using personal products with as little added scent as possible. We had to pay attention to the colors we wore – as coatis could see reds and blues. Once we found a group, we would try to stay with them, both to see what they were doing (eating, resting, moving, etc.) and to see how fast they were moving.

Carefully stalking coatis, over miles and miles of trails, game paths, and brush of the Huachucas yielded more than 3000 locations during 4½ years. Eighty-three percent of those locations were of groups, and 17% of solitary males. The number of locations per group or individual ranged from 1 to 830. I calculated home ranges using a utilization distribution model that creates a density map from the number of points in a grid overlaying the study area. Because this was a statistical model, at least 20, preferably more, locations were needed to estimate a home range.[8-12] For some of the coatis that were killed by predators shortly after being radio collared, I was not able to collect enough data to calculate a home range. Individual females were usually found with their troops except during the nesting season, so I calculated home ranges for entire troops, rather than individual females.[7] Because these models were based on individual points, they sometimes plotted separate, disconnected use areas without detecting travel in between. This is illustrated in some of the maps below.

Home range sizes in the Huachucas were some of the largest ever recorded for the species. Male home ranges averaged 610 ha (1510 ac), but there was a large span of home range sizes: 350 to 1,070 ha (865-2633 ac). Troops, exclusive of the nesting season, averaged 1360 ha (3361 ac), also with a large span 790-2240 ha (1952-5525 ac).[7] Females during the nesting season, when they were by themselves away from the troops, used much smaller areas, averaging 280 ha (692 ac). Female nesting ranges were always within the home range of their troops. Especially noticeable was that home ranges of females (in troops) were almost twice as large as those of solitary males – rare among mammals and almost unheard of among carnivorans. In the Chiricahua Mountains, 100 km NE of the Huachucas,

McColgin found slightly different results. Although the coatis she tracked also had very large home ranges (average 460 ha; 1137 ac), she found little difference in home range size between males and females.[13] However, there were methodological differences in the ways we calculated home ranges, mainly that she did not track the coatis year-round; she did not calculate home ranges separately when they were nesting versus in a group; but most importantly, what she considered separate groups, I likely would have classified as subgroups, which would have resulted in larger home range size.

Male home ranges in the Huachucas tended to follow individual canyons, whereas home ranges of the troops tended to cross several major drainages. Home ranges of troops that were lower in elevation tended to be larger. Woodlands, in general, were denser at higher elevations in the Huachucas, responding to the elevational-related increase in moisture. These woodlands also had higher soil moisture and had higher leaf-litter invertebrate populations.[14-17] The differences in sizes of home ranges related to elevation may reflect the availability of food resources.

Home range sizes in the tropics averaged less than 100 ha (247 ac) for males and slightly larger for troops.[18-21] In the tropical dry forests of western Mexico, home ranges increased to 360 ha (890 ac) during the dry season.[17] In southeastern Arizona, Shayamala Ratneyeke and colleagues. recorded home range sizes of 160-200 ha (395-494 ac) for three females during the nesting season, similar to the findings of this study of 199 ha (492 ac; using a different method of analysis: MCP).[7,22] There appeared to be a latitudinal increase in the size of home ranges, which is most likely related to food availability.

Few studies of home ranges have been done on brown-nosed coatis, but a study of one troop in southern Brazil had a home range of 540 ha (1334 ac), which shifted in location quite a bit from year to year.[14] Another study of three males in the cerrado of Tocantins, Brazil, found home ranges varied from 220 to 750 ha (544-1853 ac).[23] In Argentina, coatis had home ranges of 350-500 ha (865-1236 ac).[24] Even fewer studies have been conducted on mountain coatis, although one radiocollared male occupied 11 ha (27 ac) during a 3-month period.[25]

Home range overlap in the Huachucas was extensive, among troops, among males, and between males and troops. For males within the same canyon complex, home ranges overlapped other males by as much as 83%. Males overlapped troops by as much as 93%, and troops overlapped other

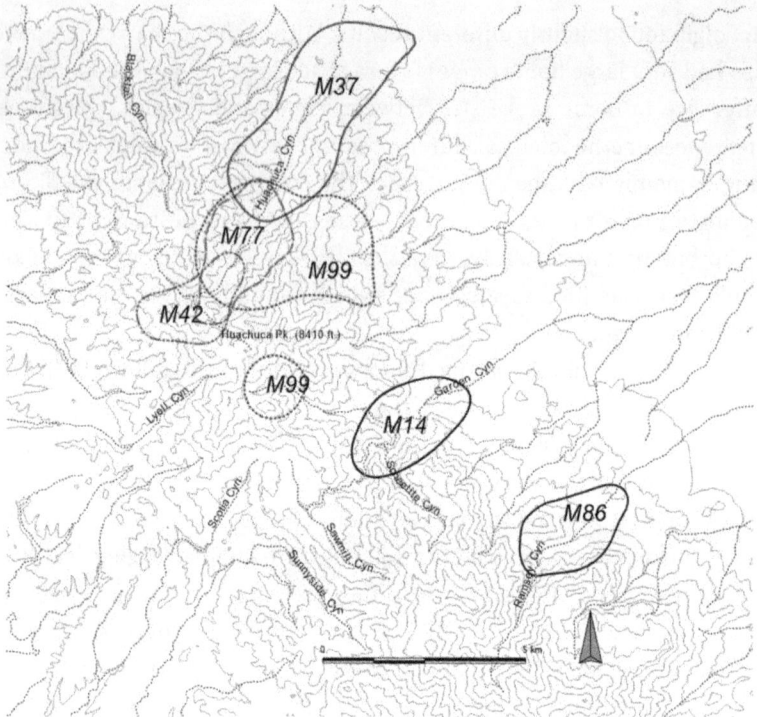

Home ranges of six male coatis in 1996. There were more coatis radioed during that year, but not enough locations were collected on them to calculate home ranges. Note that the home range of M77 is completely overlapped by other males. Note also that most of the males confine themselves to one drainage.

troops by as much as 93%. Each male's home range overlapped 2-3 troop home ranges during 1996, when density was the highest. This pattern of high overlap among home ranges and lack of territoriality also occurs in the tropics.[20,21,26] The extensive overlap of home ranges, with occasional contact among adjoining troops, makes coatis particularly vulnerable to contagious diseases. Indeed, the population in the Huachucas suffered a severe die-off from canine distemper in the early 1960's.[27]

This same pattern – of overlapping home ranges – has been found in all studies of lowland coatis. However, in areas where more home ranges are packed into an area, coati densities are much higher. For example, coati densities in the tropics were estimated at 15-56 coatis/100 ha,[21,28-30] much higher than the 1-2 coatis/100 ha found in my study and elsewhere in Arizona.[13,31,32]

The number of troops declined during the study, due to predation and drought (Chapter 4), and I was only able to keep track of two troops for the

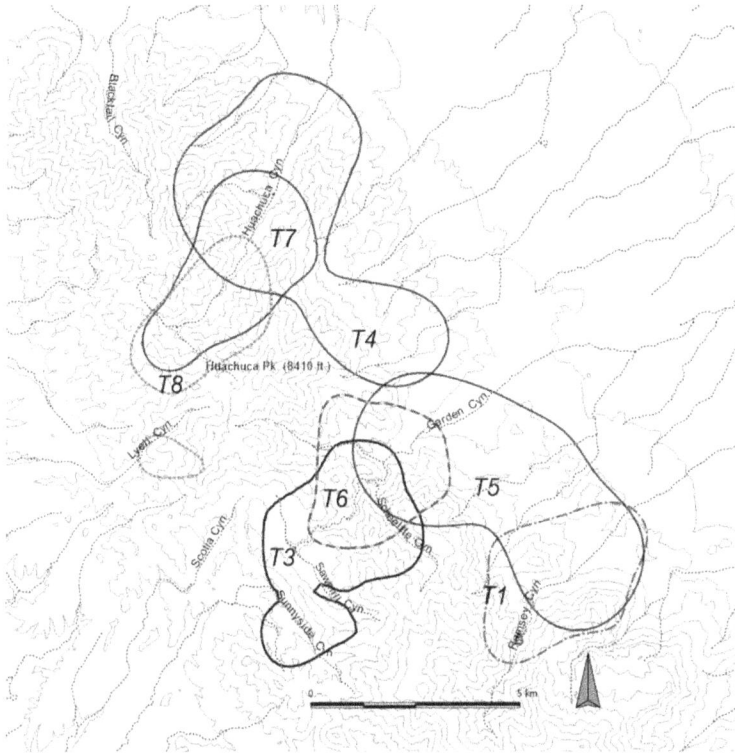

Home ranges of the troops during 1996. Compare the size and shape of the home ranges with those of the males

entire 4½ years of the study. The home ranges for those troops (Troops 3 and 5) changed a little from year to year, but both occupied their original areas. They also retained similar relationships to each other, with a slight overlap; Troop 5, which occupied lower elevations, consistently had a home range almost twice as large as Troop 3. The core areas changed a little from year to year, which I suspect was due to localized resource depletion (they ran out of food), but overall, the troops' home ranges were remarkably stable. Other troops that I monitored for only two to three years also showed fidelity to the same areas. Even as neighboring troops disappeared, the remaining troops did not move into the formerly occupied areas. Previous researchers and observers have reported that coatis in Arizona are nomadic or semi-nomadic.[33-35] My study in the Huachuca Mountains, as well as that by McColgin in the Chiricahua Mountains,[13] found that coatis are not nomadic, but have extensive home ranges, and may be absent from portions of their range for months at a time.

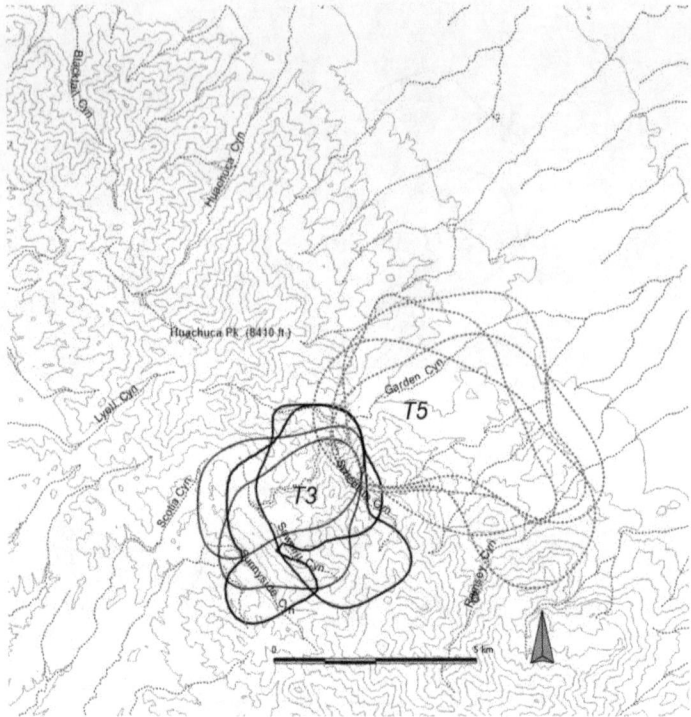

Locations of home ranges for Troops 3 and 5 for four consecutive years (1996-1999).

The advantages of stable home ranges become obvious when we look at the mortality data. During the study, predators killed 19 radiocollared coatis. I monitored 16 of these long enough to estimate their home ranges. Of those 16, 9 (56%) were killed on the peripheries of their home ranges, that is, the areas that they only used only about 5% of the time.[7] Similar studies on other species have demonstrated how important it is for animals to be intimately familiar with their surroundings and know the precise escape routes when predators appear.[36,37] If we plot the locations of coatis that were killed by predators relative to their home ranges, you can see the role of spatial familiarity. Although coatis might be killed on their periphery of their own home range, that same area was within the home range of other coatis. For example, M42 was killed within the home range of M99, and vice-versa. Likewise, two of the females from Troop 5 were killed outside of their home range but within the home range of Troop 3. In other words, there did not appear to be specific "danger zones" that all coatis avoided, but intimate knowledge of an area might include critical escape routes that occasional visitors did not have. What role perceived

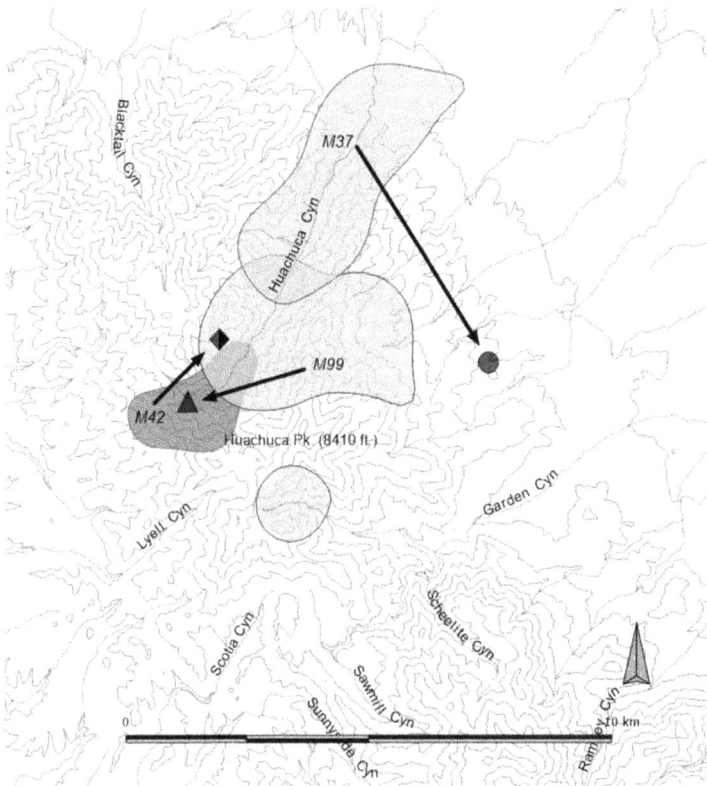

Mortalities of three males relative to their home ranges. Arrows point to locations of mortalities for each male. All mortalities in this example are due to predation by mountain lions.

predation risk had in shaping coati home ranges could not be determined in this study – it would take many more years with many more radiocollared animals to determine that.

Coatis sometimes moved long distances over the course of a day. Often, they would remain in one general area for a few days, and then move to a new area. From one day to the next, males tended to move about 862 m (0.55 mi), and females about 961 m (0.60 mi). These numbers were calculated from radio-tracking locations; in other areas where researchers have followed coatis closely, daily movements were much larger (2 to 3 km or more).[21,24] Both males and females can cross their entire home range in a day, and we saw this on few occasions. Observations and tracking coatis by looking for sign also revealed differences between the ways males and females moved through their environments. Tracks of solo males followed

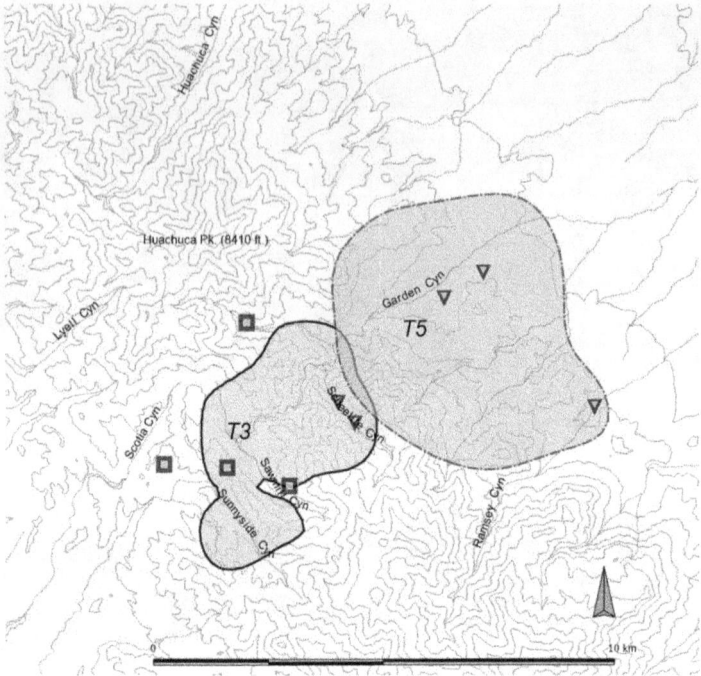

Locations of mortalities for Troops 3 (squares) and 5 (triangles) relative to their home ranges. All mortalities in this example are due to predation by mountain lions.

dirt roads for distances of a kilometer (5/8 mile) or more, and on several occasions, I saw males loping down roads and trails. Troops, on the other hand, would often stop at the edge of dirt roads and trails, and look both ways before dashing to the other side. Males also seemed more likely to become habituated to people, especially if there were handouts available.

Denning Behavior

Coatis used many night dens (to distinguish from natal, or birth, dens) throughout their home ranges. In the Huachucas, dens were usually in crevices in limestone cliffs, old mines, or sinkholes, but they occasionally constructed tree nests, too. Tree nests usually consisted of small platforms made by twisting around one or more flexible branches to form a loop. More branches were broken off and added to the nest. Nests were usually in cottonwood or ash trees, and were often near the ends of branches, rather than near the trunk. They often occurred in clusters, so that one tree may have a dozen or more nests. Some nests were built up enough to resemble hawk nests, whereas others were just a couple of

A cottonwood tree with many coati nests. San Pedro River, Arizona, 2008.

branches making a small platform. Nests were reused periodically, and it appeared that each time a nest was used, new branches were added. We also saw solo males using old hawk nests. Tree nests were common along the riparian areas in southeast Arizona and New Mexico, where rocky denning areas were harder to find. In the tropics, coatis often made large platform nests high in the canopy, these nests may be a couple of meters across.[21,38] On these larger nests, up to a dozen coatis may sleep together, with the juveniles on the inside of the huddle. Because coatis have trouble climbing trees with smooth bark, they often used vines, lianas, or neighboring trees to access preferred nesting areas.[38,39]

Habitat Use

The geographic coordinates obtained from the coati locations could be overlaid on digital habitat maps and elevation models to estimate habitat use. I determined habitat types from digital files from the Southwest GAP program,[40] and elevation and aspect of locations from a digital elevation model.

Eighty-five percent of coati locations were found in woodland habitats, specifically encinal (evergreen oak woodland), and Madrean coniferous forest. Thirteen percent of locations were found in riparian habitats, with the rest in oak savanna and urban areas. There were some differences between males and the troops: only males were found in urban areas (i.e., the housing area at Fort Huachuca), and males tended to use more riparian and encinal habitats, whereas the troops used more Madrean coniferous and oak savanna habitats, but overall differences were not significant. There were habitat differences among the troops, also, which corresponded to the elevational differences. The lower elevation troops (4 and 5) used more oak savanna, encinal, and riparian habitats, while the higher elevation troops used more Madrean conifer forest.

Elevations at coati locations differed among troops, but within each troop, elevations did not differ much by season. Only three of eight troops that were located during winter months had lower mean elevations during the winter compared with other times of the year, and these did not differ by much. In other words, most coatis did not change elevations seasonally. Both sexes of coatis were found more often on north-facing than south-facing slopes (males: 64% north-facing, 36% south-facing; females: 62% and 38%). However, both sexes increased their use of south-facing slopes during the winter, although the difference was significant only for females.[41]

We found a greater diversity of habitats used by coatis than noted by previous researchers in Arizona. Most of them recognized the coati's affinity for woodland habitats, although they were based on sightings and may have been biased toward areas most accessible by roads and trails. Wallmo and Gallizioli recorded 72% of coati observations in riparian deciduous habitat and oak-pine woodland in the Huachucas.[35] In the Chiricahua Mountains, Dirk Lanning found coatis often on brushy canyon slopes and described their habitat as evergreen woodland between 1400 and 2000 m.[31] Kaufmann and colleagues summarized coati habitat in the southwest as primarily encinal and pine-oak woodland, with occasional sightings in montane conifer forest, riparian deciduous forest and Sonoran desertscrub.[34]

Resident coatis also inhabit cottonwood-lined watercourses, often at some distance from encinal or pine-oak associations, such as Upper Cienega Creek, Aravaipa Creek, and the San Pedro River. Studies of food habits and movements of these coatis are lacking. In addition, more than a dozen solo coatis have been seen over the years at Organ Pipe National Monument, in the Sonoran Desert.[34] It is important here to distinguish between the presence of solo males and troops. Solo males are often seen far from any known troop (including the sightings in Organ Pipe), especially during the mating season but also at other times of the year. These are most likely occasional wanderers and not indicative of a local breeding population. Because of the vulnerabilities of small groups (Chapter 4), at least four adult females are needed to be form a sustainable troop, and thus a breeding population.

Habitats used by both white-nosed and brown-nosed coatis in the tropics were comparable to those in the Huachucas. Coatis in those areas preferred woodland, but also used riparian forests and savannas. Urban and grassland habitats were used rarely.[18,26,42-44] One study in the Atlantic forests of Brazil found the coatis were more arboreal than in other studies, which the researchers attributed to the coatis foraging on the insects found within the bromeliads high in the canopy.[45]

Because it made seeing the coatis so much more difficult while radio-tracking, we were frustratingly aware that they liked dense habitats that provided both horizontal (side) and vertical (overhead) cover. Although the species composition of habitats used in Arizona is not comparable to species used in the tropics, all species of coatis use dense, wooded or brushy habitats.[46,47] Presumably coatis seek these habitats not only for their food sources, but also to hide from predators.

Notes and References

1. Burt WH (1943) Territoriality and home range concepts as applied to mammals. Journal of Mammalogy 24:346–352

2. McNaughton BL, Battaglia FP, Jensen O, Moser EI, Moser M-B (2006) Path integration and the neural basis of the "cognitive map." Nature Reviews Neuroscience 7:663–678

3. Powell RA (2000) Animal home ranges and territories and home range estimators. *In*: Boitani L, Fuller TK (eds) Research techniques in animal ecology: controversies and consequences. Columbia University Press, New York, NY, pp 65–110

4. Spencer WD (2012) Home ranges and the value of spatial information. Journal of Mammalogy 93:929–947

5. Ripple WJ, Beschta RL (2004) Wolves and the ecology of fear: can predation risk structure ecosystems? BioScience 54:755–766

6. Tolon V, Dray S, Loison A, Zeileis A, Fischer CA, Baubet E (2009) Responding to spatial and temporal variations in predation risk: space use of a game

species in a changing landscape of fear. Canadian Journal of Zoology 87:1129–1137

7. Hass CC (2002) Home-range dynamics of white-nosed coatis in southeastern Arizona. Journal of Mammalogy 83:934–946

8. Size of home range was calculated using a fixed kernel density estimator (Worton, 1995; Seaman and Powell, 1996). Following convention in defining home range as that area normally covered by an animal (Burt 1943); the outer 5% of locations were removed from analyses (White and Garrott, 1990). Estimators were calculated using Ranges V software (Kenward and Hodder, 1996) and Manifold GIS. The default smoothing factor from Ranges V, calculated as the standard deviation divided by the sixth root of N, where N is the number of locations (Kenward and Hodder, 1996).

9. Worton BJ (1995) Using Monte Carlo simulation to evaluate Kernel-based home range estimators. Journal of Wildlife Management 59:794–800

10. Seaman DE, Powell RA (1996) An evaluation of the accuracy of kernel density estimators for home range analysis. Ecology 77:2075–2085

11. White GC, Garrott RA (1990) Analysis of wildlife radio-tracking data. Academic Press, New York, NY

12. Kenward RE, Hodder KH (1996) Ranges V: An analysis system for biological location data. NERC, Wareham, Dorset, UK

13. McColgin ME, Koprowski JL, Waser PM (2018) White-nosed coatis in Arizona: tropical carnivores in a temperate environment. Journal of Mammalogy 99:64–74

14. Beisiegel B de M, Montovani W (2010) Habitat use, home range and foraging preferences of the coati *Nasua nasua* in a pluvial tropical Atlantic forest area. Journal of Zoology 269:77–87

15. Alves-Costa CP, da Fonseca GAB, Christofaro C (2004) Variation in the diet of the brown-nosed coati (*Nasua nasua*) in southeastern Brazil. Journal of Mammalogy 85:478–482

16. Levings SC, Windsor DM (1982) Seasonal and annual variation in litter arthropod populations. *In:* Leigh EG Jr (ed) The ecology of a tropical forest. Seasonal rhythms and long-term changes. Smithsonian Institution Press, Washington, D.C., pp 355–387

17. Valenzuela D, MacDonald DW (2002) Home-range use by white-nosed coatis (*Nasua narica*): limited water and a test of the resource dispersion hypothesis. Journal of Zoology 258:247–256

18. Caso A (1994) Home range and habitat use of three neotropical carnivores in northeast Mexico (*Felis pardalis, Felis yagouaroundi, Nasua narica*). Master's Thesis, Texas A&M, Kingsville

19. Estrada A, Halffter G, Coates-Estrada R, Meritt DA Jr (1993) Dung beetle attracted to mammalian herbivore (*Alouatta palliata*) and omnivore (*Nasua narica*) dung in the tropical rain forest of Los Tuxtlas, Mexico. Journal of Tropical Ecology 9:45–54

20. Gompper ME (1997) Population ecology of the white-nosed coati (*Nasua narica*) on Barro Colorado Island, Panama. Journal of Zoology, London 241:441–455

21. Kaufmann JH (1962) Ecology and social behavior of the coati, *Nasua narica*, on Barro Colorado Island, Panama. University of California Publications in Zoology 60:95–222

22. Ratnayeke S, Bixler A, Gittleman JL (1994) Home range movements of solitary, reproductive female coatis, *Nasua narica*, in south-eastern Arizona. Journal of Zoology, London 223:322–326

23. Trovati RG, Alves de Brito B, Duarte JMB (2010) Habitat use and home range of brown-nosed coati, *Nasua nasua* (Carnivora: Procyonidae) in the Brazilian Cerrado biome. Revista de Biologia Tropicale 58:1069–1077

24. Ben Hirsch, personal communication.

25. Rodriguez-Bolanos A, Sanchez P, Cadena A (2003) Patterns of activity and home range of mountain coati, *Nasuella olivacea*. Small Carnivore Conservation 29:16–19

26. Valenzuela D, Ceballos G (2000) Habitat selection, home range, and activity of the white-nosed coati (*Nasua narica*) in a Mexican tropical dry forest. Journal of Mammalogy 81:810–819

27. Risser SC Jr (1963) A study of the coati mundi *Nasua narica* in southern Arizona. Master's Thesis, University of Arizona, Tucson

28. Glanz WE (1991) Mammalian densities at protected versus hunted sites in central Panama. *In*: Robinson JG (ed) Neotropical wildlife use and conservation. University of Chicago Press, Chicago, IL, pp 163–173

29. Gompper ME (1994) The importance of ecology, behavior, and genetics in the maintenance of coati (*Nasua narica*) social structure. Ph.D. Dissertation, University of Tennessee, Knoxville

30. Wright SJ, Gompper ME, DeLeon B (1994) Are large predators keystone species in Neotropical forests? The evidence from Barro Colorado Island. Oikos 71:279–294

31. Lanning DV (1976) Density and movements of the coati in Arizona. Journal of Mammalogy 57:609–611

32. Hass CC (1997) Ecology of white-nosed coatis in the Huachuca Mountains, Arizona, a preliminary study. Final report submitted to the Arizona Game & Fish Department, Phoenix, AZ. 1–52

33. Pratt JJ (1962) Establishment and trends of coati-mundi in the Huachucas. Modern Game Breeding 23:10–11, 15

34. Kaufmann JH, Lanning DV, Poole S (1976) Current status and distribution of the coati in the United States. Journal of Mammalogy 57:621–637

35. Wallmo OC, Gallizioli S (1954) Status of the coati in Arizona. Journal of Mammalogy 35:48–54

36. Forrester TD, Casady DS, Wittmer HU (2015) Home sweet home: fitness consequences of site familiarity in female black-tailed deer. Behavioral Ecology and Sociobiology 69:603–612

37. Metzgar LH (1967) An experimental comparison of screech owl predation on resident and transient white-footed mice (*Peromyscus leucopus*). Journal of Mammalogy 48:387–392

38. Olifiers N, Bianchi RC, Mourao G, Gompper ME (2009) Construction of arboreal nests by brown-nosed coatis, *Nasua nasua* (Carnivora: Procyonidae) in the Brasilian Pantanal. Zoologia 26:571–574

39. Personal observation.

40. Southwest Regional GAP. http://swregap.nmsu.edu/. Accessed 29 Jul 2018

41. Oneway ANOVA of aspect (degrees) by calendar quarter, Males: $F_{3,522}$ = 1.18, P = 0.3184; Females: $F_{3,2508}$ = 3.78, P = 0.0108.

42. Alves-Costa CP, Eterovick PC (2007) Seed dispersal services by coatis (*Nasua nasua*, Procyonidae) and their redundancy with other frugivores in southeastern Brazil. Acta Oecologica 32:77–92

43. Costa E, Mauro R, Silva J (2009) Group composition and activity patterns of brown-nosed coatis in savanna fragments, Mato Grosso do sul, Brazil. Brazilian Journal of Biology 69:985–991

44. Desbiez ALJ, Borges PAL (2010) Density, habitat selection and observations of South American coati *Nasua nasua* in the central region of the Brazilian Pantanal wetland. Small Carnivore Conservation 42:14–18

45. Beisegel B de M (2001) Notes on the coati, *Nasua nasua* (Carnivora: Procyonidae) in an Atlantic forest area. Brazilian Journal of Biology 61:698–692

46. Gompper ME, Decker DM (1998) Nasua nasua. Mammalian Species 580:1–9

47. Gompper ME (1995) Nasua narica. Mammalian Species 487:1–10

15

The Mating Game

The Mating Season Begins

When coati mating season begins in the Huachucas, the hot, dry spring has just started. It is mid-March, and the Madrean oaks are beginning to change color and drop their leaves. If it has been a particularly dry winter, the leaves will not reappear until the summer rains start, and for 2-3 months the area looks more like winter in the southeastern U.S. than like spring in the west. In wetter years, reddish-brown leaves are quickly followed by soft leaves of lime green as the forest goes through a quick color change. Unlike deciduous trees in much of the temperate zone that drop their leaves to protect them from freezing temperatures, Madrean deciduous trees drop their leaves as an adaptation to seasonal drought. This occurs throughout the thorn-scrub and tropical deciduous forests of Mexico, Central, and South America.[1] As spring approaches, daytime temperatures rapidly climb above 32°C (90°F). The streams start drying up and water is limited to springs and artificial reservoirs. Near the springs and in the shadiest spots in the mountains, a few flowers bloom; one of the few real indications of spring. The real flush in plant growth will begin late June or early July with the arrival of the rains.

Last years' youngsters are now eight months old and are starting to lose their high-ranking status in the troop. In dry winters, when food is scarcer, overwinter survival of youngsters can be quite low, so the large troops observed the previous fall are much smaller now. As mating season approaches, there is an increased tendency for groups to break into sub-groups and start moving about their home ranges in search of mating opportunities. More recent research in a variety of populations has found many differences in coati mating patterns. In this chapter I will describe the coati mating behavior in the Huachucas, compare it to other populations, and discuss the similarities and differences.

Mating Systems Defined

The mating behavior of most mammals falls into four broad categories: monogamy, where one male and one female maintain an exclusive bond for at least one breeding season; polygyny, where one male mates with multiple females; polyandry, where one female mates with multiple males; and promiscuity, where males and females each have multiple partners and no lasting bonds.[2] Monogamy is common among birds, as it usually takes two adults to take care of the eggs and feed the nestlings. Among mammals, monogamy is quite rare, and genetic analyses have revealed that even what looks like monogamy often is not. Only 10% of mammals maintain long-term monogamous relationships.[3] Polygyny is the most common mating system among mammals.[2] In both polygyny and polyandry, some bonding occurs, resulting in exclusive access to individuals of the opposite sex during the mating period, and sometimes extending into the period of raising the juveniles. In promiscuity, no such bonds occur.

The goal of this all, of course, is to get the greatest number of one's genes into future populations, what scientists term "fitness." Mammals have developed many strategies for propagating their genes, and some of these result in conflicts of interest between the sexes. Males want to get their sperm into as many good mothers as possible. In some species, to ensure the survival of their offspring (or to make sure the offspring are indeed theirs), putative fathers instinctively stay with the female and help rear the young. In other cases, the females are capable of rearing young on their own or with a group of other females, so the males are free to try to inseminate as many females as possible.

Females, on the other hand, are physiologically limited in the number of offspring they can produce, so they may be more interested in quality rather than quantity. If they can choose (and not all do), they seek out traits that might reflect "good" genes. They not only want to produce strong healthy offspring, but they need to make sure the male's genes are compatible with their own genes (think of the Rh blood factor in humans). Ideally their offspring would be not only good survivors, but also good subsequent breeders.

For males, obtaining mating opportunities often means some form of mate guarding. The tactics males use to monopolize females, and thus mating opportunities, depends on how females are distributed in space and time, which is a function of resource distribution, predation pressure, and the activity of other males. Different tactics include defending individual females during their receptive period (tending); defending feeding territories that overlap female home ranges (territory defense); defending a group of females for most or part of the year (harem defense); or defending specific mating territories within a communal mating ground (lekking). Usually each male is on his own, but in some mammals, such as African lions, adult males cooperate in obtaining mating opportunities.[3] Sometimes tactics are combined, for instance, a male may defend a territory, then maintain a harem of females within his territory.[2] Other combinations have been observed, so it's difficult to pigeon-hole all animals into specific mate-guarding systems.

Mating systems are not a characteristic of a species, rather they are the outcomes of decisions by individuals of a species, and they may differ among populations or among individuals of a population.[2,3] For example, we normally think of humans as monogamous, but there exist cultures where polygyny is the norm, and others where polyandry and promiscuity are widely practiced. It is not unusual for animals to switch strategies as they age, or even within a mating season.[4] For instance, while studying pronghorn in Montana, John Byers found that they shifted from a system of territorial defense to one of harem defense. This occurred after an extremely cold winter killed most of the old males. This shift appeared to be a function of male age, as younger males were less likely to set up territories. However, several years later, when the males should have shifted to a territorial defense system, males were still practicing harem defense, indicating a role of experience. *"Males appear to adjust their probability of defending a territory on the basis of the frequency with which they encounter other territorial males."*[4]

Usually it is the older, higher ranking males that can defend individual females or groups of females from other suitors. The ability to reach and hold high status is often appealing to females. However, the younger and/or lower ranking males seldom sit on the sidelines – the stakes of missed opportunities are too high. So lower ranking males are on the constant lookout for opportunities, either by constantly pressuring the guarding male, hoping to overturn his throne, or trying to steal away some females within the guarding male's harem. Sometimes the best strategy is to lurk in the shadows, and hope the females come to you. Males that hover around harems or pairs are referred to as "satellites" whereas males that search a wide area looking for mating opportunities are called "rovers" or "opportunists."

Likewise, the females have strategies of their own to meet their goals of strong, healthy offspring. Because there are often lower ranking males nearby, or high-ranking males in the next territory over, females may have the choice of accepting the male that holds their harem or territory or sneaking away. Sometimes they do both. Usually, the ability to hold high rank and guard multiple females is correlated with genetic quality, so these males tend to get most of the mating opportunities and father the most offspring.

Mating Systems in Coatis

From Jason's field notes [my comments in brackets]:

> *April 6, 1997, 1400 h. F104's and M042's troop observed at loc #14 on Huach. Pk NE map [this was our way of marking locations where we observed or detected coatis], only 2 coatis seen; both unmarked. Watched from 1400-1410, both sunbathing/resting on rocky cliff, one approx. 6 ft below the other. At 1410, lower one moved up to other coati and "nudged" it. Within 1 minute they were mating!! Mating for exactly one hour (1410-1510); pair never separated during that time. Male (who was slightly darker) held female around the waist the whole time. Female seemed relaxed throughout. Male would thrust vigorously for 1-2 seconds, rest for up to 1 minute (all the while clinging to the female), then repeat. While I watched, I could hear coati chuckling [probably what we now call barking] lower down the mtn, presumable at me. I explored these sounds twice (both F104 and M042's signals were strong in that direction) but saw no other coatis. As soon as I got back to my vantage point, I heard the chuckling again. At 1510, female stood and moved away from male. I got the impression that*

"parting" may have been painful for her, as she made a sound almost like a snort, then high-pitched squeaks, then ran off down the other side of rock. Male stood around for about 20 sec, then walked away.

This observation of Jason's turned out to be an important moment in the coati study. I would make a similar observation a year later.[5] These observations were quite different from those made by John Kaufmann in Panama. Kaufmann had described a system in which one adult male was allowed entrance into the troop just before the mating season. Presumably the male mated with all the females, although Kaufmann observed only a few copulations.[6] All of the copulations Kaufmann observed were only a few seconds long, and when I described Jason's observation to him, he was surprised. Our observations were similar to those Gilbert had made years earlier in the Huachucas,[7] but were the disparities due to real differences between coatis in Panama versus coatis in Arizona, or are things a lot more complicated than previously described?

What Kaufmann described for coatis at his study site in Panama was a harem defense mating system. One high-ranking male carefully earned the good graces of the troop females and became a loosely integrated member of the troop. He regularly groomed and was groomed by the adult females, and occasionally played with the juveniles and subadults. This arrangement usually extended for some weeks before and after the actual mating season. As mating season approached, he vigorously fought off any approaching males.

Kaufmann noted an increase in vocalizing by the males, both chuckling and chop-chop sounds. Males also increased their rates of urine-rubbing, especially after agonistic encounters with other males or adult females.[6] He noticed an increase in mounting behavior by subadult males and a sudden increase in nocturnal activity. By flashlight, coatis were seen chasing each other through the trees, vocalizing loudly, fighting, and attempting copulation. *"It was clear that the disturbances were caused by adult males approaching and chasing unreceptive band members."*[6] So it appears nearby satellite or roving males were trying to usurp females from the harem holder. Kaufmann focused his observations on the troops, so it was not always clear what the solitary males were up to.

There were differences in the level of acceptance into the troops by individual males. Females, in general, would not allow a male to join a troop unless he spent some time becoming familiar to them. This often included staying on the periphery for a while, and carefully making advances to start grooming. Even so, some males were only allowed to hang around the periphery of the group, and other males were allowed to mix freely.[6,8] The males remained subordinate to the females and would quickly retreat if a female lunged or chittered at them.

Courtship among coatis is a brief and casual affair. There appear to be no preliminaries or elaborate courtship displays. No flowers, no chocolate, no displays of affection. The male walks up to the female (or vice-versa), nudges her, mounts, and begins thrusting. If the female is not receptive, the male is met with squeals and a face full of sharp teeth. If she is receptive, she stands passively while the male thrusts. Both Jason and I observed copulations of 60 minutes, and several other researchers have also noted prolonged copulations of 30 minutes or more.[5,7-9] The end of the long copulation appears to be painful to the females, as they were observed squealing and running away from the male.[5] Hirsch observed one female coati immediately bite a male after mating, leaving a large bloody wound.[10] It is not known if the length of copulation is correlated in any way with the quantity or quality of sperm transfer.

As mentioned in Chapter 5, all births occur within a 1- to 2-week period after a 70-77-day gestation, so all the females are in estrus for a very brief period. Males and females have a limited time to acquire mating opportunities.

Gilbert's observations of one troop in the Huachucas were similar to Kaufmann's, with a few notable differences. All the females in one troop plus several older males gathered around the cabin he used as a research station. The highest-ranking male associated with the troop for most of the mating season. He actively defended the troop against other adult males. But if the male left, one of the other males quickly joined the troop. He saw the high-ranking male copulate with at least four different females, and one of those four females also copulated with one of the other males. All of the copulations were brief, less than five seconds. He did see the high-ranking male leave the troop accompanied by a female on two occasions (two different females). The couples remained away from the troop for an hour or more in each instance. So, this also appears to be a harem defense mating system, although a bit looser than what Kaufmann described in Panama. Gilbert and his crew noticed a large increase in vocalizing during the mating period, particularly chuckling. Gilbert routinely put out food for the coatis, including large amounts of dog food and marshmallows. This probably influenced the spatial distribution of that group of coatis, although it appears to have had little detrimental impact, as all the adult females became pregnant.[7]

Booth-Binczik's observations of mating behavior among the white-nosed coatis in Guatemala were completely different than anything described previously. Rather than one male associating with a troop, she saw several males accompanying troops. The males would climb trees near the foraging group and start calling. Females would leave the troop and join the male in his tree to mate. Copulations were 20-60 minutes long, although some were in progress when first seen. She described the vocalization as like an alarm bark. She described this as a "mobile lek," in which males set up tiny mating territories (a tree), and females choose among territory holders for mating. As the troop moved on to forage in new areas, males would descend from their trees, catch up with the troop and find new trees from which to call.[8,11] Males defended their perches from other males. What factors females used to choose among mates was not clear, although Booth-Binczik hypothesized that scent marks might be involved.[8,12] Lekking is rare among mammals, so her observations in Guatemala are very interesting (and controversial).

The population of coatis that Booth-Binczik observed in Guatemala differed in several ways from other studied groups. The group sizes were exceptionally large, often more than 50 animals. Males often associated with the troops outside of the mating season, with individual adult males found with troops about half of the time, and troops having one or more males nearby in 2/3 of encounters. Males associated with several different troops, and each troop had multiple males associated with it. The males were usually peripheral and not an integrated part of the troop. Booth-Binczik suggested that one of the reasons that males were associating with the troops was to develop relationships with the females. As with Kaufmann's coatis, she felt that familiarity might be an important factor when females chose a mate.[8]

Hirsch's observations of brown-nosed coatis in Brazil also found a harem-satellite mating system. In that population, adult males were integrated into troops year-round. During the mating season, up to five satellite males followed each troop. These males appear to stay with the same troop for the duration of the mating period. The harem males also actively tried to keep the satellite males out of the troop, spending much of each day chasing and fighting. But all the effort appeared to be worth it: the harem males were much more successful at obtaining mating opportunities than individual satellite males.[13]

What we saw in the Huachucas during my study could not really be classified as a harem system or mobile lekking. Usually when we found troops during the mating season, they did not have males with them. Likewise, the marked males were seen by themselves most of the time. In each mating season (1996-2000), and overall, in only 27% of observations of radiocollared males were they associating with a troop. Similarly, only 33% of observations of troops during the mating season included adult males. And rarely was the same male seen more than once with a troop, with a few exceptions. But most of the times those troops were seen during the mating season, males were not present. Males were seen visiting up to three different troops, and troops were seen with up to three different males, including unmarked males.

The observation Jason made of mating coatis implies a satellite system, but it is unclear as to whether there was a harem male with the troop or not. We knew that M42 was periodically associating with the troop, but Jason could not get close enough to see whether he was in the troop or just nearby. The observation I made of coatis mating in 1998 was similar. In that instance, I had found Troop 5 within their home range, which included radiocollared females F149 and F252 and at least half a dozen unmarked animals, including an unmarked adult male. They started to move up and over a ridge, and I followed a short distance behind them, not getting too close so I would not disturb them. As I scanned the frequencies of the radiocollared females, I noticed that F149 had stopped while the rest of the troop kept heading up the hill. I moved in a little closer and saw that she was copulating with an unmarked male. I hunkered down behind a bush and watched and took notes. The mating behavior was nearly identical to what Jason had described a year earlier. Twenty minutes after I first saw the coatis copulating, the other radiocollared female in the troop (F252) approached the pair and sat down next to them, just a few inches away. As soon as he detected the other female, the male started barking and making chop-chop vocalizations and my impression was that he was concerned about being attacked by the second female. After a couple of minutes, F252 left and moved out of sight, and the male quieted down and continued copulating.

At just over an hour, I could hear chirp-grunts from coatis in the canyon below me, and the copulating male began barking again. I soon saw several coatis moving toward the pair, and the male and female quickly separated with loud squeals (I could not tell who was vocalizing). F149 ran toward the approaching group, and the radio receiver told me that F252 was with that group, which must have circled around us to get back to the canyon bottom. The male that had copulated with F149 headed up the ridge in the opposite direction, just as a large male left the coati group and went after him. The male from the troop chased the other male up and over the ridge.

It was difficult to tell exactly what was going on with the group dynamics (other than the obvious mating). It appeared that the copulating male was not the male I had seen with the troop, but another male that had been nearby, possibly a satellite. It also appeared that F252 left the copulating pair, went back to the troop, and brought them back to where the pair was, possibly to disrupt the copulation. That may be reading too much into the situation, but I did get the impression that she wanted to copulate with the same male as F149, but she could not if he was busy with F149. Bringing the troop over, especially the male associating with them, effectively freed the satellite male to mate again (provided he could physically do so after an hour of vigorous copulation).

Based on the studies to date, it appears that coatis have a promiscuous mating system, usually, but not always, characterized as a harem-satellite system. Males defend groups of females, but not territories. Variants include the observations of the coatis in the Huachucas (temporary harem-satellite and/or roving) and the observations in Guatemala (mobile lekking). The different populations studied to date differ in density and sex ratios, and the propensity of males to associate with troops outside of the mating season (which may be a function of density and sex ratio). The strategies used by individual males may differ depending on the density and distribution of troops and males, which are a function of habitat productivity. In other words, in tropical habitats that support higher densities of coatis, a male coati may be more successful if he can remain with one harem for the duration of the mating period for his group. If female group sizes get too large for a single male to defend, something like a mobile lekking system develops. I assume the weak harem system seen in the Huachucas relates to the low density (20x lower than some tropical populations), but it is not clear why a male would choose to leave a troop to seek other mating opportunities.

Paternity

Gompper tested the genetics of the white-nosed coatis he was studying in Panama, probable descendants of the populations studied by Kaufmann, using paternity exclusion tests to figure out if the harem holder was indeed fathering most of the offspring in the troop. And the results were quite surprising: the harem holders sired very few of the offspring of the females that they guarded. Although his sample sizes were small, it was clear that the offspring of any given troop were fathered by multiple males, not just the harem holder. And in one troop, the harem holder sired none of the tested offspring.[14] So the females were obviously sneaking off for dalliances with males outside of the troop, begging the question of why a male would bother to go through all of the fights with other males and efforts to appease the females in the troop, if it did not increase his reproductive success?

As part of her study of a large population of white-nosed coatis in Guatemala, Booth-Binczik examined patterns of paternity among one large troop. She also found that many fathers were contributing to the troop's offspring. In addition, she found that more than half of the litters she sampled had multiple fathers within the same litter. This meant that individual females were mating with more than one male while they were in estrus. Four of the litters she sampled each had three different fathers.[8] Hirsch found similar results for brown-nosed coatis in Argentina. In that population, an adult male was present in each troop year-round. Paternity testing of 59 offspring from three groups found that roughly half of litters had some offspring fathered by satellite males, and they fathered 1-2 offspring per litter. The majority of offspring were fathered by the group male.[13]

Every genetic study of coatis has found that the offspring in a troop, in any given year, have multiple fathers. In addition, some females are copulating with more than one male, so that their litters also have multiple fathers. Other carnivorans, including raccoons, European badgers, and mink also show multiple paternity within litters.[15-17] So why are females mating with males other than the harem master, and why are they mating with multiple males? No studies to date have presented data on the age or other circumstances of the female coatis that mated with multiple males to determine what factors might be correlated with multiple paternity.

Studies on other animals may give us insights into possible benefits to females of carefully choosing the quality and quantity of potential mates.

Given the tight synchrony of estrus among females, harem males must copulate rapidly in succession. It is possible that some sperm depletion occurs, so females may be ensuring that enough sperm are able to fertilize all her eggs by mating with multiple males.[8] Likewise, she may be able to choose males that are more genetically compatible. Several studies that have looked at multiple paternity in other animals have found that females that mated with more than one male were more likely to conceive and give birth, and also weaned larger litters.[18-22] Mating with more than one male at or near the time of ovulation sets the stage for sperm competition,[23] which may mean that the most vigorous sperm are the ones that fertilize the eggs. Too few studies have been conducted in coatis to determine if females or their offspring really benefit from mating with multiple males.

Males often set up home ranges that overlap their maternal home ranges, so females risk mating with sons or brothers if they mate with the males with whom they are most familiar. Inbreeding is known to cause many problems with conception and offspring viability. But in the populations that have looked at the genetics of the males that join troops, in nearly every case, they males are not related to the troop they join.[13,14] Mating with satellite males may be a way for females to further insure against inbreeding, or perhaps to balance inbreeding and outbreeding.[24]

Coatis on the Move

What was very noticeable during the mating season in the Huachucas was how much and how far the coatis moved. Coatis dramatically increased their ranging movements beginning in mid-March. Both males and troops went wandering far beyond their home range boundaries in search of mates; roughly half of the locations of both troops and males were outside of the home ranges they had occupied for the previous 6-9 months. This is likely greatly underestimated, as the radiocollared coatis became much more difficult to find during the mating season, and it was common to lose touch with a troop or male for a week or more, before they would suddenly reappear in the middle of their home range. Both males and females increased their rates of scent marking, and it is possible they were leaving scent trails to attract the opposite sex.

Agonistic Behavior

Fights among male coatis can be very vicious, and some of the worst fights occur during the mating season. At the end of the season, most males bear deep wounds – gashes, torn claws, even teeth ripped out. Although I saw the results of this fighting in the captured males, I did not see any actual fights. However, the casualties of some of the fights might be informative.

On April 1, 1998, I located M77 with Troop 9 in part of their home range. It was a bit north of M77's home range, but within the boundaries of Troop 9's home range. Prior to that, I had located M77 within his normal home range, but Troop 9 had wandered widely both northwest and south of their usual area. I did not see any of the animals, but the signals were coming from the same location, so I assumed M77 was with the troop. On April 2nd, I located M77's signal by itself up a little side canyon from where he was found the day before, and Troop 9 was a little further north, but within their home range. On April 3rd, M77 had not moved from his location, and Troop 9 was a little further east. I snuck up on the troop, and saw most of the troop, plus a large male. Over the next few days, Troop 9 moved several canyons to the north, but M77 stayed in the same spot.

On April 6th, I climbed up to M77's location. I found him hiding under a small rock overhang more than halfway up the mountain. He was severely injured; one front leg was very swollen, and he was unable to put any weight on it. His coat was very rough, and he had other cuts and gashes. He appeared to be in a lot of pain and did not alarm bark or try to flee when I approached. I watched him for a few minutes, gave him the apple from my lunch, and then left, not wanting to disturb him any more than necessary. I checked him again on April 8th. He was close to his little shelter, but he was actively foraging. He was still unable to put any weight on his front leg. He remained near that spot for several more days, before moving back to the center of his home range. He limped on that front leg for at least six months, at which point the battery in his radio collar died and I lost contact with him.

We saw similar circumstances with M14 in 1996, when he engaged in a fight severe enough to curtail any further mating activities (see Chapter 13). Fights this severe have also been observed in other populations, although of the few studies that have been conducted, the harem masters remained with their harems throughout the mating season, with little to no turnover.[6,8,13] There is obviously a lot we do not understand about what factors determine the various tactics used by individual males and female coatis.

Mortality Rates

The predation rate on male coatis increased during the mating season, perhaps due to venturing outside of their home ranges, injuries sustained during fights, or perhaps they became less cautious around predators around that time. Four of seven radiocollared males that were killed by predators died during March and April, while only 1 of 12 radiocollared females killed by predators died during the same period (most females died during the nesting period). Keeping up with coatis that were moving much more than normal, and recovering mortalities meant the mating season was exhausting for the researchers, too.

And I will never forget the spring of 1997 when, in addition to being fortunate enough to see the above mating interaction, I also had to recover two mortalities in the same day. On 16 April, I picked up a mortality signal from F100, a member of Troop 7, in Huachuca Canyon. I climbed up the steep slope on the south side of the canyon and found her remains near a rock outcrop about a third of the way up the slope. The radio, and pile of hair and bone chips were consistent with a mountain lion kill and appeared to have been there for a couple of days. Before heading to the bottom of the canyon, I used the telemetry receiver to scan for M42, whose signal I had picked up earlier in the day. This time, I picked up his signal across the canyon from me, only now it was in mortality mode (pulsing twice as fast as normal). So, I descended the 100 m (300 ft) to the bottom of the canyon and started up the other side.

Chasing down a mortality signal can be very time consuming, as it usually involves moving forward a short distance, scanning for the signal to make sure you have not passed it, then moving toward the signal again. In this case, it also involved scrambling through scrubby oak and pinyon trees, and up a very steep and slippery scree slope until I finally found M42's remains hidden under some low branches – 300 m (1000 ft) above the canyon bottom. After crawling through the bushes to recover what was left of M42 (hair and bone fragments, again consistent with a mountain lion kill), I slipped and slid my way down the steep slope, only to find myself at the edge of a huge catclaw acacia thicket at the canyon bottom. Too tired to climb up and around the thicket, I forced my way through, which was not a good idea as I was wearing shorts. By the time I got through the thicket, I felt like I had been stung by 100 bees, and my legs and arms were bleeding from dozens of small cuts. I made it back to my truck as the sun set behind the mountain. But my day was not over yet – I still had traps to check in Garden Canyon. It was one of the very few times I really hoped there were no coatis in the traps, and luckily, there were not.

Beyond the Mating Season

The coati mating season in the Huachucas winds down in mid-April. Virtually every adult female is now pregnant, and faces the challenge of not only feeding herself, but a rapidly growing litter of baby coatis, and doing so at a time when food is becoming scarce. The coatis return to their home ranges and recover from the wild activity of the mating season. The variations in behavior between populations tell us is that coatis are highly adaptable, and capable of adjusting their behavior to the current circumstances. Too few populations have been studied for adequate time for us to really know what kind of mating system will prevail in an area, what strategy a particular male might use to obtain mating opportunities, or what factors females might be considering when they choose a mate.

And this brings us back to where we started – with the birth of the new coatis.

Notes and References

1. Murphy PG, Lugo AE (1986) Ecology of tropical dry forest. Annual Review of Ecology and Systematics 17:67–88

2. Brown JL (1975) The Evolution of behavior. W. W. Norton & Company, New York, NY

3. Clutton-Brock TH (1989) Mammalian mating systems. Proceedings of the Royal Society of London 236:339–372

4. Byers JA (1998) American pronghorn: social adaptations and the ghosts of predators past. University of Chicago Press, Chicago, IL

5. Hass CC, Roback JF (2000) Copulatory behavior of white-nosed coatis. The Southwestern Naturalist 45:329–331

6. Kaufmann JH (1962) Ecology and social behavior of the coati, *Nasua narica*, on Barro Colorado Island, Panama. University of California Publications in Zoology 60:95–222

7. Gilbert B (1973) Chulo. A year among the coatimundis. Alfred A. Knopf, New York, NY

8. Booth-Binczik SD (2001) Ecology of coati social behavior in Tikal National Park, Guatemala. Ph.D. Dissertation, University of Florida, Gainesville

9. Smith HJ (1980) Behavior of the coati (*Nasua narica*) in captivity. Carnivore 3:88–136

10. Ben Hirsch, personal communication.

11. Booth-Binczik SD, Binczik GA, Labisky RF (2004) Lek-like mating in white-nosed coatis (*Nasua narica*): socio-ecological correlates of intraspecific variability in mating systems. Journal of Zoology 262:179–185

12. Rich TJ, Hurst JL (1998) Scent marks as reliable signals of the competitive ability of mates. Animal Behaviour 56:727–735

13. Hirsch BT, Maldonado JE (2011) Familiarity breeds progeny: sociality increases reproductive success in adult male ring-tailed coatis (*Nasua nasua*). Molecular Ecology 20:409–419

14. Gompper ME, Gittleman JL, Wayne RK (1997) Genetic relatedness, coalitions and social behaviour of white-nosed coatis, *Nasua narica*. Animal Behaviour 53:781–797

15. Hauver S, Hirsch BT, Prange S, Dubach J, Gehrt SD (2013) Age, but not sex or genetic relatedness, shapes raccoon dominance patterns. Ethology 119:769–778

16. Dugdale HL, Macdonald DW, Pope LC, Burke T (2007) Polygynandry, extra-group paternity and multiple-paternity litters in European badger (*Meles meles*) social groups. Molecular Ecology 16:5294–5306

17. Yamaguchi N, Sarno RJ, Johnson WE, O'Brien SJ, Macdonald DW (2004) Multiple paternity and reproductive tactics of free-ranging American minks, *Mustela vison*. Journal of Mammalogy 85:432–439

18. Hoogland JL (2013) Why do female prairie dogs copulate with more than one male? Insights from long-term research. Journal of Mammalogy 94:731–744

19. Løvlie H, Gillingham MAF, Worley K, Pizzari T, Richardson DS (2013) Cryptic female choice favours sperm from major histocompatibility complex-dissimilar males. Proceedings of the Royal Society of London B: Biological Sciences 280:20131296

20. Firman RC, Simmons LW (2008) Polyandry, sperm competition, and reproductive success in mice. Behavioral Ecology 19:695–702

21. Fox CW, Rauter CM (2003) Bet-hedging and the evolution of multiple mating. Evolutionary Ecology Research 5:273–286

22. Stockley P (2003) Female multiple mating behaviour, early reproductive failure and litter size variation in mammals. Proceedings of the Royal Society B: Biological Sciences 270:271–278

23. Birkhead TR, Hunter FM (1990) Mechanisms of sperm competition. Trends in Ecology & Evolution 5:48–52

24. Bateson P (1983) Mate choice. Cambridge University Press, Cambridge, UK

PART THREE
THE HUMAN
CONNECTION

16

Coatis Past

A s an ecologist, I try to unravel the complex interactions between an animal and its environment - where it lives, what it eats, what eats it, and so on, and try to understand the role of place in that animal's ecology. In other words, how much does where it is living influence its ecology, and how do things like food habits and reproduction vary in different parts of an animal's range. So far, that has been the focus of this book. But the world is an ever-changing place and conditions now are not what they used to be. To understand more about coati ecology at the northern end of their range and their ability to adapt to a changing world, we need to go back in time.

Early Records of Coatis

Coatis are reportedly recent immigrants to the southwestern U.S. The first coatis were recorded in what is now the United States near Brownsville, Texas in 1877. At the time, it was common for baby coatis to be sold near the U.S.-Mexican border as pets, and this first sighting was probably an escaped pet. Most, if not all, subsequent sightings in Texas have resulted from released or escaped pets, and there are no known natural breeding populations.[1] At the time of that first sighting, Arizona and New Mexico were still territories and neither would achieve statehood until 1912. Even though the area had been occupied by indigenous peoples for more than 10,000 years, and by Europeans since the mid-1700s, detailed scientific accounts of flora and fauna only date back 150 years.

Following the Treaty of Guadalupe Hidalgo (1848), which ceded much of New Mexico Territory (including what is now northern Arizona) to the United States, and the Gadsden Purchase (1853) which added southern Arizona and southwestern New Mexico, the U.S. government sent survey crews out to describe and document the geology, flora, and fauna of the newly acquired territory.[2] No coatis were recorded during these surveys. Nor were they recorded during surveys in the late 1800s.[3-5] Other than the early sightings in Texas, coatis did not appear in what is now the United States until 1892, when a couple of coatis were observed on Fort Huachuca (established in 1877). One of these was collected and sent to the U.S. National Museum.[6] The border was a contentious area until well after 1900, getting dragged into battles as part of the Mexican Revolution and American Indian wars, which no doubt dampened the enthusiasm of biologists to do inventories. Coatis were first formally recognized as a component of the fauna of the United States in 1934, with several coatis collected from the Huachuca, Patagonia, and Santa Rita Mountains, and one observed as far north as the Blue Range of Arizona, some 150 miles north of the border.[7] By the late 1940s, they were frequently seen, although I hesitate to use the word "common," in the Huachuca and Patagonia Mountains.[8]

Why should we care when coatis arrived in the southwestern United States? In addition to presenting an interesting ecological question (why then?), it also brings up questions related to management – should they be treated as an invasive species or part of the native fauna? I suspect that coatis have inhabited the border region for hundreds, perhaps thousands, of years prior to it becoming part of the United States. According to research on marked animals, including my studies, coatis disperse and colonize new areas very slowly, so their reported rapid expansion into mountain ranges throughout southeastern Arizona and southwestern New Mexico is quite surprising. Within 30 years of their discovery, they occupied much of the available habitat south of the Gila River. Despite their noisy, gregarious nature, coatis can be remarkably difficult to find when their numbers are low. I suspect, as do John Kaufmann and others,[9] that numbers were low enough during the boundary surveys that they were simply missed.

But what about archaeological evidence? The area north of the current U.S.-Mexico border has attracted many archaeologists interested in the rich cultural legacies left by the Ancestral Puebloans (previously referred to as the Anasazi) and their contemporaries. I contacted several archaeologists specializing in the fauna of southern and central Arizona. Most were unaware of any archaeological evidence of coatis north of the U.S.-Mexico border. Anne Woosley, Director of the Amerind Foundation, qualified those missing observations:

> No site collections from SE Arizona, SW New Mexico, or northern Mexico in which coatis have been identified immediately come to my mind...In my experience, many faunal assemblages from earlier excavated sites contain mis-identified materials. For example, antelope is identified as gracile doe.[10]

Archaeologist Will Russell expanded on this thought in an email:

> Coati obviously could have lived here, but I don't know why their remains would not be found in excavations, especially given the large number of digs going on at any given time. Most archaeologists are not faunal experts (myself included) and merely bag animal bones for later analysis. There is undoubtedly a backlog of such bones waiting to be analyzed, but even still... Could their bones be mistaken for those of raccoon?[11]

I did find one report that mentioned coati remains found in a rock overhang on Fort Huachuca (the site of my studies), but there were some concerns with temporal contamination (the date the bones were deposited could not be determined accurately) and correct identification. If the bones were actually coati and consistent with the strata they were found in, then they would have been at least 500 years old.[12]

My search of the FAUNMAP and Paleobiology databases, which record fossil specimens, and the scientific literature turned up a variety of specimens from across North America of coati ancestors – Arctonasua and Paranasua – from millions of years ago.[13] Several extinct species of coatis were also found, including Nasua pronarica in Texas[14] and Nasua mastodonta in Florida.[15] But more interesting, one specimen of Nasua narica (white-nosed coati) was identified from Murrah Cave near Del Rio, Texas, from the Late Holocene (4500-450 years before present).[16]

I looked up the reference for the Murrah Cave excavation. Coati bones were found in a cave that had been occupied by hunter-gathers for long periods. Other species found in the cave included bison, raccoons, jackrabbits, cottontails, ringtails, badgers, kit foxes, coyotes, wolf, deer, pronghorn, ocelot, and a variety of fish, shellfish, and birds. There was no evidence of domesticated crops such as maize, squash, or melons, indicating that the remains in the cave may have predated the arrival of agriculture in the southwest, at least 3,000 years ago.[16] The publication did not describe what bones were found, or how many individuals they represented. There may be other archaeological evidence of coatis north of the border in the forms of pottery art and rock art (petroglyphs and pictographs).

Early Inhabitants of the U.S.-Mexico Borderlands

Roughly 1000 years ago, plus or minus a few hundred years, what is now the southwestern United States was occupied by several indigenous peoples - the Hohokam, the Patayan, the Fremont, the Mogollon, the Ancestral Puebloans. They occupied much of what is now Arizona and New Mexico, but their extensive trade networks reached from Tenochtitlan (now Mexico City) north through the Colorado Plateau and likely through the entire southwest.[17] They were descendants of the original hunter-gatherers that once roamed the area, feeding on mammoths and giant ancient bison. Several thousand years ago, they took up farming by adapting plants domesticated in southern Mexico to drier conditions. Maize (corn) was domesticated more than 7,000 years ago in what is now Oaxaca in southern Mexico.[18] It would eventually spread north and east, where Pilgrims found Native Americans growing corn when they arrived on the east coast of the U.S. in 1620. Maize also spread into Central and South America and would be taken up by predecessors to the Inca.[19]

The importance of the arrival of agriculture to southwestern and Mesoamerican cultures cannot be overstated. In addition to providing us with crops like corn, squash, beans, tomatoes, chiles, and chocolate, the development of a stable source of food had profound impacts on Mesoamerican societies. Without the daily requirement of hunting and gathering food, populations could increase, and sophisticated cultures could develop. The tremendous cities built by the Mesoamericans, and eventually the Incans, were tied directly to the development of agriculture. The arrival of maize, beans, and squash in the southwest also allowed the expansion of the Hohokam and Ancestral Pueblo cultures, and to a lesser extent the Patayan and Mogollon. Large Pueblo cities arose out of dust in the southwest. Extraordinary buildings, usually constructed of local adobe, rose several stories high.

These cultures of the southwest were trade partners to peoples as far away as the Valley of Mexico, which in turn traded with peoples as far away as the Yucatan Peninsula and Central America. So, during the peak of these cultures, from around 800 to 1450 AD (depending on the culture), there were well-developed connections from northern Arizona and New Mexico all the way to Central America, and possibly even further south.[20] Similar cultures arose in South America at the same time, with the Wari (Huari) and Tiwanaku occupying much of Peru, Bolivia and northern Chile. The presence of maize at archaeological sites in Bolivia is evidence of trade and cultural transmission between peoples of Central America and these cultures.[21]

Many of these cultures kept animals as pets. The peoples that crossed the Bering Land Bridge during the Pleistocene brought their dogs with them. Dogs were companions, guards, pack, and draft animals, hunting assistants, and sometimes food. Some cultures, including the Ancestral Puebloans, buried dogs with their owners, showing the esteem with which the animals were held. Macaws were also kept as pets, and Mesoamericans probably kept monkeys, coatis, raccoons and other animals as pets, although they haven't been found in burial sites.[22] Even today, coatis are kept as pets throughout this region.[23,24] Macaws were traded from central Mexico to at least as far north as Chaco Canyon in northern New Mexico,[18] and genetic and archaeological evidence indicates that turtles were traded throughout Central America for food or ritual.[25] Could coatis also have been traded as pets or a source of food? Although they have not been recorded from burial sites (which may be due to their absence, misidentification, or being among materials that have not been analyzed yet), I believe there is evidence that the peoples of the southwest 1000 years ago were familiar with coatis.

Coati remains have been recovered from archaeological sites of pre-Columbian Maya, most notably on the island of Cozumel. Mayans may have imported coatis from the mainland more than 1000 years ago, although limited recent genetic analyses indicate that the coatis predate the Mayans on the island by a few millennia.[26] Mayan women on Cozumel Island had a special relationship with coatis both as pets and for food, and they may have symbolized fertility. As Bishop Landa wrote around 1566, *"The Indian women raise them, and they leave nothing which they do not root* over and turn upside down, and it is an incredible thing how wonderfully fond they are of playing with the Indian women and how they clean them from lice."[23] Coatis were one of the most numerous animals represented in remains found on Cozumel, although few coati remains have been found at other Mayan sites.[27] However, coatis are one of many animals, including monkeys, kinkajous, and parrots, that have been kept as pets among the Maya for hundreds of years, and likely long before that.[22,23] According to Brand, coatis appear in the Popol Vuh[28] as teasers of the twin gods, and may appear in at least one Mayan codex.[23] Coatis are also commonly represented on prehistoric ceramics.[23]

Coati Depictions on Pottery

Some of the most beautiful images of coati on pottery come from the southwest U.S. The most common and dramatic examples come from the funereal bowls of the Mimbres peoples that used to live near what is now the Gila Wilderness in southwestern New Mexico. The Mimbres were a subculture of the Mogollon people, one of a group of peoples that occupied the southwest a millennium ago. The Mimbres painted bowls with incredibly artistic, and often life-like, images of people and animals. What exactly these images depict - things actually seen or fantasized - is unknown, but some of the images are incredibly detailed and accurate, and often depict the local food animals.[29] The bowls were placed over the face or head of the deceased, after a hole was punched in the center of the bowl to allow the spirit access to the next world.

The first coati-like image I saw was one of the ones on the next page. It was described in several sources as a mountain lion, based on the white-tipped tail (where this attribution came from, I do not know, as mountain lions do not have white-tipped tails). However, it is a nearly perfect image of a coati: a long snout with elongated lower canines, a long body with relatively short legs, an exceptionally long tail, a mask on the face, small, rounded ears, and five toes on each foot (mountain lions only have four toes on each hind foot). As I began to research Mimbres pots, I found quite a few more images, usually identified as mountain lions or "unknown quadruped," that appeared to be coatis.[30]

I am convinced that the Mimbres people were familiar with coatis, although I am not sure if it was because coatis were present in the area or due to trade with peoples further south. As archaeologist Will Russell describes:

Examples of images on Mimbres funereal bowls that may illustrate coatis. Left, a Mimbres bowl that depicts a very accurate coati, although it was listed as a mountain lion.[30] Sketch from photographs and drawings.[29] Right, drawing of a bowl identified as "unknown quadruped eating juniper berries."[30] Juniper berries are one of the coatis preferred food in this area. Designs were on the inside of the bowls.

> *I have seen some Mimbres bowls which I would agree were coati. The problem with Mimbres bowls is that there are some with such detailed representations of marine fish that they can be ID'd to the species level and having come from the Sea of Cortez. We're only now starting to realize how truly mobile prehistoric populations were. It's been my experience (albeit limited) that the more portable an item is, the more likely the iconography has been 'borrowed' out of its original context. That's totally anecdotal. But I can't think of too many examples of rock art depicting animals which weren't there. We never find buffalo motifs, for instance, in the desert.[11]*

More examples of coatis on pottery are found from many locations in southern Mexico and Central America as well as the southwestern U.S. Coati images in pottery may represent something metaphorical, rather than literal.[29] If coatis represented anything metaphorical, in my opinion, it would likely have been related to fertility, family, or perhaps community. This metaphor could explain the Isle of Coati, in Lake Titicaca in Bolivia. During the Incan Empire, in the 1400s, Lake Titicaca was the site of two major sacred islands - the Island of the Sun, associated with male rulers, and the Island of the Moon, also known as the Isle of Coati, which may have been a nunnery. Both islands figure prominently in the creation mythology of the Inca.[31] The ruins on both islands predate the Incan

Left, ceramic whistle in the shape of a woman with a coati in her lap. The pottery dated from 300-600 CE and was found in Veracruz. Sketch from photograph at online auction site. Right, pottery recovered at Chaco Canyon, New Mexico, that is believed to represent a coati. National Park Service photo.

Empire by thousands of years and appear to be Tiwanaku holy sites. Various scholars have tried to explain the use of the term "coati," with none attributing it to the species, although the term was derived from Tupi Indians who likely had trade networks with the Tiwanaku. Some scholars attribute the word as "coata," which is part of the local Quechua dialect for "queen."[32] Coatis, mountain or brown-nosed, are few to non-existent in that region of Peru and Bolivia today, although they may have occurred there in the past. There is at least one pottery image of a possible coati attributed to the Inca.

Coati Depictions in Rock Art

After leaving coati project, Jason Roback received a grant from New Mexico Game & Fish Department to survey coatis in the southwestern part of the state. On one of his travels, he stopped by the Three Rivers Petroglyph Site in south-central New Mexico. This was an area previously inhabited by Jornada Mogollon - neighbors of the Mimbres. Jason thought some of the rock art images at Three Rivers looked like coatis. As with the Mimbres pottery images, most of these have been interpreted as mountain lions. It is much more difficult to assign species to rock art due to the coarseness of the images. I visited Three Rivers a short time after Jason made me aware of it and agreed that some images could be interpreted as coatis.

Chimu-Inca ceramic possibly representing a coati. Date between 1400 and 1533. Photo by Luis Garcia, licensed under Creative Commons 3.0 (from Wikimedia commons.)

Examining hundreds of petroglyphs at Three Rivers made me stop and assess what a coati image in rock art would look like. If I lived 1000 or more years ago, and wanted to depict a coati in a petroglyph, how would I do it? What are the critical elements that would make it a coati, and not something else? I think it would need a long nose and long tail. Stubby ears, short legs, flat feet and five claws per foot - but maybe claws would be too tedious to add. Some of these features, such as a long tail and stubby ears, are also common with the big cats (mountain lions and jaguars). If I wanted to portray coatis, I might include multiples - of different sizes - to reflect their social nature. So, with this image in my head, I started exploring known rock art sites, looking for possible coati images.

I began searching sites in southeastern Arizona and contacting other people for sightings. The search was slow until I signed up for a rock art walk at Patagonia Lake State Park, near the U.S.-Mexico border in Arizona. We found some interesting glyphs at the park, the most interesting being one that might be interpreted as a monkey or a more creative interpretation of a coati. Monkeys are not found within hundreds of miles of the area, making the glyph even more interesting. Another rock on a nearby hill also included a group of petroglyphs that might be coatis, and several

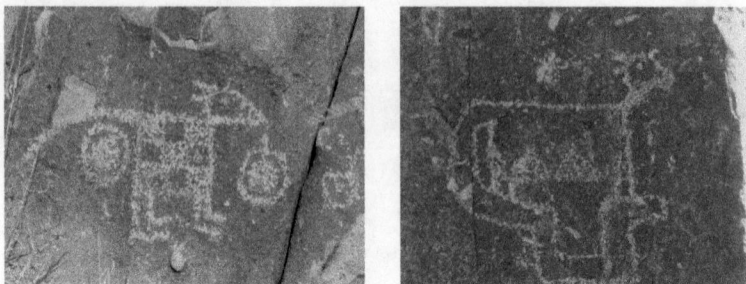

Three Rivers petroglyphs that may represent coatis.

miles downstream from Patagonia Lake was a pictograph that appeared very coati-like.

I started searching in earnest for coati images in rock art. My search would take me to many sites from Texas to Nevada, and northern Mexico. I also scoured books on rock art of the southwest (it would take many lifetimes to examine all the rock art left behind by ancient peoples of the southwest). I found a good number of images that fit my mental picture of what a pecked or painted coati might look like. The clear majority of these were found within the range of the southwestern farmers, in what is now coati habitat.

Interpretations of Rock Art

Only the artists who created these images know what they really represent. Interpreting rock art can be a controversial, sometimes even contentious, endeavor. Theories for their interpretation range from fantasy images induced during trances, to clan symbols, to hunting mojo, to doodles. My guess, given that these images were created over thousands of years, is that all these theories apply depending upon which image you are looking at. I believe that ancient peoples probably created rock art for some of the same reasons modern people create what we often refer to as graffiti: personal expression, declarations of turf, affiliation with a group. Additionally, ancient peoples may have created rock art to document hunting accomplishments or increase hunting success. An increasing popular idea among archaeologists is that images depicted in rock art represent metaphorical images derived from shamanic vision quests.[33,34] However, Ekkehart Malotki takes a more inclusive approach, recognizing that production of rock art may serve religious, utilitarian, as well as purely decorative functions, and may include doodling of both children and adults.[35]

A pictograph (painted image) that might be interpreted as a monkey, or maybe a coati. Some of the other images on the rock might be interpreted as coatis.

In his research at Hohokam sites near Phoenix, archaeologist Will Russell found evidence of a coati clan among the proto-Hopi, similar to presumed coati clan symbols in northern Arizona. (The Hopi are descendants of the Ancestral Puebloans and perhaps other southwestern farmers). He described the complexity of these relationships:

> *The Hopi have no Coati (or similar) Clan currently or within memory of oral tradition, but do acknowledge that there may have once been one. Some Hopi consider the Mimbres as an ancestral component, so the presence of coati on Mogollon bowls made sense to me in that regard. The Hopi believe that many of their clans originated to the south at a land called Palatkwapi. No one knows where this was, for sure. I've made the argument that it was along the lower Salt River. This is why I wasn't terribly surprised to find coati petroglyphs here. So far, we have found four which I feel comfortable identifying as coati. Interestingly, two occur one above the other, with the top one inverted, much like many Mimbres Style III bowl designs. All can be tentatively dated to the Preclassic Hohokam era (ca. AD 475-1150), with the most convincing one likely originating in the Colonial or Sedentary period (ca. AD 750-1150).[11]*

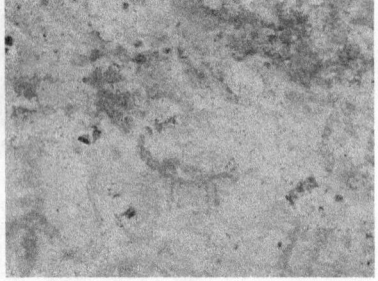

Petroglyphs and a pictograph near Patagonia Lake, Arizona, that might be interpreted as coatis.

Why didn't more archaeologists identify coatis in archaeological sites? Most of the archaeologists that conducted the first site surveys around the turn of the last century were from universities in the eastern U.S.[18] It is possible that they did not know what coatis were (as coatis were not considered part of the U.S. fauna yet) and may not have made the connections between the extensive trade with southern Mexico. Little archaeological work was being done just across the border in Sonora or Chihuahua. In other words, they did not see evidence of coatis because coatis were outside their scope of knowledge. In addition, basic information on coati natural history was lacking until the 1950s, and detailed studies of coati distribution at the northern end of their range would not be conducted until the 1970s. There is still a paucity of archaeological and natural history information on coatis in northern Mexico.

I think it is safe to say that some of the rock art and pottery images represent coatis and that, therefore, some of the southwestern farmers were familiar with them. But the question remains as to whether coatis lived throughout the area or were traded as pets and/or food. Could coatis have lived in the area 1000 years ago, or were they just a well-traveled cultural metaphor?

Climatic Changes

According to paleoclimate research in the southwest, the climate has been rather dynamic since the glaciers melted, with periods of both warming and cooling. Although the overall trend has been warming, resulting in many plant and animal species shifting their ranges northward and/or higher in elevation, at various times it was wetter, cooler, warmer, and drier than today.[36] From approximately 2500 to 1000 years before present

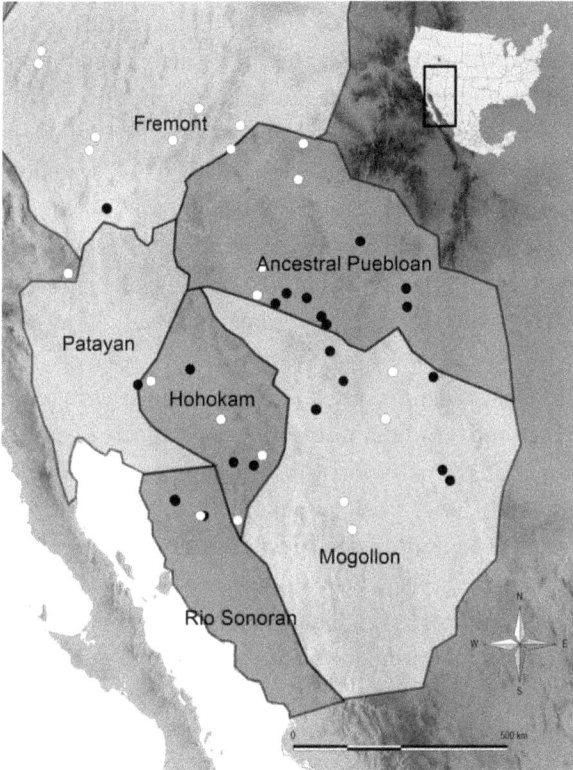

Map of rock art sites visited or identified through publications that contained images that might be interpreted as coatis (black circles) or no such image (white circles). Estimation of cultural boundaries from Cordell and MeBrinn.[18]

(depending on which model and what part of the globe you look at) the earth experienced a warming period. In the southwestern U.S., much of this time was also wetter.[37-39] This corresponded with the expansion of several cultures, including the agricultural peoples.[36,40] Widespread agriculture during this time reflects an increased amount and predictability of summer rains - an expansion northward of the North American Monsoon.[18] Outside of the predictable areas of summer rains, peoples such as the Fremont and Cosos depended more upon hunting and gathering than agriculture.[41] A long period of warm and relatively wet weather characterized the Medieval Warm Period (950-1250 AD); cultures of the time responded to the conditions by expanding agriculture, creating larger pueblos, and greatly increasing population sizes.

This was followed by a long cooling trend that started about 1250 AD, which decreased summer rains at the time when populations had significantly expanded and become dependent upon agricultural crops.[42,43] Several decades of drought, especially lack of summer rain, forced peoples to move closer to reliable rivers and establish irrigation systems.[44] As we've seen lately in the southwest, long-term droughts are also accompanied by catastrophic wildfires, which can convert large areas of forest to grasslands or shrublands; this also would have affected people's ability to support themselves on the land by removing fuel wood and building materials.

These climate changes resulted in major changes in the distribution of peoples in the southwest and would also have affected coatis. Coatis are very dependent on the monsoon rains and their impacts on insect populations. The expanded human populations that occurred prior to the droughts would have impacted local wildlife populations, through habitat change and hunting, especially of animals considered as food.[45] So, based on the cumulative impact of climate change and overuse of local resources by indigenous peoples, I would expect the coati population to decline after 1250 AD. Temperatures continued to decline until around 1700 AD, with a noticeable increase in temperatures beginning in the late 1800s. The expansion of coati populations in the last century coincides with warming temperatures.

So, it is likely that the area could have supported coatis 1,000 years ago and that coatis were known to the human cultures around at that time. But it is also possible that coatis were traded, physically or metaphorically. Perhaps coatis' northward movement after the retreat of the glaciers was aided by humans. Coatis in Arizona have close genetic ties to those in northern Mexico,[46] so if they were traded, they didn't travel very far. The answer may lie in the many boxes and bags of skeletal material collected from archaeological digs waiting for analysis or may lie hidden in the memories of the ghosts of peoples who lived in the southwest so long ago. Hopefully, I have brought awareness to the issue, so southwestern archaeologists consider coatis as part of the fauna present during this time, instead of dismissing them outright.

Notes and References

1. Taber FW (1940) Range of the coati in the United States. Journal of Mammalogy 21:11–14

2. Emory WH, Baird SF, Conrad TA, Englemann G, Hall J, Parry CC, Schott A, Torrey J, Girard C (1857) Report on the United States and Mexican boundary survey made under the direction of the Secretary of the Interior. C. Wendell, printer, Washington, D.C.

3. Allen JA (1895) On a collection of mammals from Arizona and Mexico, made by Mr. W.W. Price, with field notes by the collector. Bulletin of the American Museum of Natural History 7:193-258

4. Allen JA (1906) Mammals from the states of Sinaloa and Jalisco, Mexico, collected by J.H. Batty during 1904 and 1905. Bulletin of the American Museum of Natural History 22:191-262

5. Mearns EA (1907) Mammals of the Mexican Boundary of the United States. Smithsonian Institution, Washington, D.C.

6. Pratt JJ (1962) Establishment and trends of coati-mundi in the Huachucas. Modern Game Breeding 23:10-11, 15

7. Taylor WP (1934) Coati added to the list of United States mammals. Journal of Mammalogy 15:317-318

8. Wallmo OC, Gallizioli S (1954) Status of the coati in Arizona. Journal of Mammalogy 35:48-54

9. Kaufmann JH, Lanning DV, Poole S (1976) Current status and distribution of the coati in the United States. Journal of Mammalogy 57:621-637

10. Anne Woosley, personal correspondence, 1999.

11. Will Russell, personal communication.

12. Altschul JH, Cottrell MG, Clement M, Towner, Ronald H (1993) The Garden Canyon Project: studies at two rock shelters, Fort Huachuca, southeastern Arizona. Statistical Research Inc., Tucson, AZ

13. Baskin JA (1982) Tertiary Procyoninae (Mammalia: Carnivora) of North America. Journal of Vertebrate Paleontology 2:71-93

14. Webb SD, Hulbert RC Jr, Morgan GS, Evans HF (2008) Terrestrial mammals of the Palmetto fauna (early Pliocene, latest Hemphillian) from central Florida phosphate district. Natural History Museum of Los Angeles County 41:293-312

15. Emmert LG, Short RA (2018) Three new Procyonids (Mammalia, Carnivora) from the Blancan of Florida. Bulletin of the Florida Museum of Natural History 55:17

16. Holden WC (1937) Excavation of Murrah Cave. Bulletin of the Texas Archaeological and Paleontological Society 9:48-73

17. I recognize my geographic bias here - the borderlands region is part of the southwestern US but is part of northwestern Mexico.

18. Cordell LS, McBrinn ME (2012) Archaeology of the Southwest., 3rd edition. Left Coast Press, Walnut Creek, CA

19. Bruno MC (2008) Waranq waranq: ethnobotanical perspectives on agricultural intensification in the Lake Titicaca basin (Taraco Peninsula, Bolivia). Ph.D. Dissertation, Washington University, St. Louis, MO

20. Sutherland K (1998) MesoAmerican ceremony among the prehistoric Jornada Mogollon. *In*: Smith-Savage S (ed) Rock art of the Chihuahuan desert

borderlands. Sul Ross State University and Texas Parks and Wildlife Department, pp 61–87

21. Ramirez R, Timothy DH, Diaz E, Grant UJ (1960) Races of maize in Bolivia. National Academy of Sciences - National Research Council, Washington, D.C.

22. Morley SG, Brainerd GW (1983) The ancient Maya. Stanford University Press, Stanford, CA

23. Brand DD (1964) The coati or pisote (*Nasua narica*) in the archaeology and ethnology of Meso-America. *In:* Sobretiro del XXXV Internacional de Americanistas: Mexico. Mexico City, pp 193–202

24. Leopold AS (1959) Wildlife of Mexico. The game birds and mammals. University of California Press, Berkeley

25. González-Porter GP, Maldonado JE, Flores-Villela O, Vogt RC, Janke A, Fleischer RC, Hailer F (2013) Cryptic population structuring and the role of the Isthmus of Tehuantepec as a gene flow barrier in the critically endangered Central American river turtle. PLOS ONE 8:e71668

26. McFadden KW, García-Vasco D, Cuarón AD, Valenzuela-Galván D, Medellín RA, Gompper ME (2010) Vulnerable island carnivores: the endangered endemic dwarf procyonids from Cozumel Island. Biodiversity and Conservation 19:491–502

27. Hamblin NL (1984) Animal use by the Cozumel Maya. University of Arizona Press, Tucson, AZ

28. The sacred book revered as a source of ancient Mayan culture, traditions, beliefs, and history.

29. Fewkes JW, Brody JJ (1989) The Mimbres. Art and archaeology. Avanyu Publishing, Inc., Albuquerque, NM

30. Cunkle JR (2000) Mimbres mythology: tales from the painted clay. Golden West Publishers, Phoenix, AZ

31. Arkush E (2005) Inca ceremonial sites in the southwest Titicaca Basin. *In:* Stanish C, Cohen AB, Aldendorfer MS (eds) Advances in Titicaca Basin archaeology. Cotsen Institute of Archaeology, UCLA, Los Angeles, CA, pp 209–242

32. Bauer BS, Stanish C (2010) Ritual and pilgrimage in the ancient Andes: the islands of the Sun and the Moon. University of Texas Press, Austin, TX

33. Cunkle JR, Jacquemain MA (1995) Stone magic of the Ancients: petroglyphs, shamanic shrine sites, ancient Rituals. Golden West Publishers, Phoenix, AZ

34. Janine Hernbrode, personal communication.

35. Malotki E (2007) The rock art of Arizona: art for life's sake. Kiva Publishing, Walnut, CA

36. Polyak VJ, Cokendolpher JC, Norton RA, Asmerom Y (2001) Wetter and cooler late Holocene climate in the southwestern United States from mites preserved in stalagmites. Geology 29:643–646

37. Ely LL, Enzel Y, Baker VR, Cayan DR (1993) A 5000-Year Record of Extreme Floods and Climate Change in the Southwestern United States. Science 262:410–412

38. Grimm EC, Lozano-García S, Behling H, Markgraf V (2001) Holocene Vegetation and Climate Variability in the Americas. *In:* Interhemispheric Climate Linkages. Academic Press, Inc., Cambridge, MA, pp 325–370

39. Toomey RS, Blum MD, Valastro S (1993) Late Quaternary climates and environments of the Edwards Plateau, Texas. Global and Planetary Change 7:299–320

40. Polyak VJ (2001) Late Holocene climate and cultural changes in the southwestern United States. Science 294:148–151

41. https://en.wikipedia.org/wiki/Fremont_culture, accessed 8/27/2015.

42. Benson LV, Berry MS, Jolie EA, Spangler JD, Stahle DW, Hattori EM (2007) Possible impacts of early-11th-, middle-12th-, and late-13th-century droughts on western Native Americans and the Mississippian Cahokians. Quaternary Science Reviews 26:336–350

43. Betancourt JL, Van Devender TR, Martin PS (1990) Packrat middens: the last 40,000 years of biotic change. University of Arizona Press, Tucson, AZ

44. Childs C (2008) House of rain: tracking a vanished civilization across the American Southwest. Back Bay Books, New York, NY

45. Cannon MD (2000) Large mammal relative abundance in Pithouse and Pueblo Period archaeofaunas from southwestern New Mexico: resource depression among the Mimbres-Mogollon? Journal of Anthropological Archaeology 19:317–347

46. Nigenda-Morales SF, Gompper ME, Valenzuela-Galván D, et al (2019) Phylogeographic and diversification patterns of the white-nosed coati (*Nasua narica*): Evidence for south-to-north colonization of North America. Molecular Phylogenetics and Evolution 131:149–163

17

Coatis Future

T he preceding chapters have described coati biology, behavior, and population dynamics, and prehistoric and historic distribution. What does the future hold for coatis, in an era of warming climates, disappearing forests, and increasing human pressure? Coatis are widely spread throughout the tropical Americas. They occupy most forested or brushy habitats. They are generally considered common throughout their ranges, although in some areas they appear to be declining due to loss of habitat, overhunting, and possibly genetic isolation.[1]

Current Status in the Americas

The International Union for the Conservation of Nature (IUCN) classifies both white-nosed and brown-nosed coatis as "Least Concern," although it qualifies that with:

The numbers of this species [referring to the white-nosed coati] are
unknown and population estimates range from rare to common. It is
rare in the United States and can be anything from common to scarce
in Central America where its status is less well known, but indications
are that its numbers have been greatly reduced...The Mexican
population has probably been severely reduced and may even be
extirpated in certain areas. Population density is greater in the tropics

231

> *than in southwestern United States. Both regions show year-to-year*
> *fluctuations in population sizes as a result of disease or food*
> *availability.[1]*

The population of coatis on Cozumel Island appears to be critically Endangered,[2] but taxonomic confusion as to whether it is a separate species (*N. nelsoni*) or a subspecies of the white-nosed coati has precluded listing it as Endangered by IUCN. For brown-nosed coatis, the IUCN finds,

> *This species is listed as Least Concern because it is widespread and*
> *apparently common in an area of largely intact habitat, population*
> *density varies greatly from region to region and there are no major*
> *threats (although the species is probably declining locally through*
> *hunting and habitat loss).[3]*

Western mountain coatis are considered Threatened in Columbia and Ecuador due to recent population declines related to extensive habitat loss.[4] Their status in Peru is unknown, with only a few specimens reported[5,6] and even less known about their distribution, numbers, or ecology. Eastern Mountain coatis are considered Endangered, due to both limited distribution and habitat loss,[7] although their status as a distinct species has been questioned (see Chapter 1). CITES (Convention on International Trade in Endangered Species of Wild Fauna and Flora), which regulates trade in live wildlife and animal parts, lists white-nosed coatis in Honduras and brown-nosed coatis in Uruguay in Appendix III, reflecting concerns in international trade of these species at sufficient levels to impact populations.[8] Much of this trade has been baby coatis for the pet trade.

In the United States, coatis appear to be expanding their range in Arizona and New Mexico,[9,10] although as mentioned in the previous chapter, this may be coatis repatriating ranges that they occupied a thousand years ago. Throughout their range in the U.S., their densities are low, and they are extremely vulnerable to periodic die-offs due to drought, predation (as seen in my research), and disease.

It is difficult to measure the impacts of habitat loss, export for the pet trade, and hunting, on coati populations because they are so hard to count. Even though they are diurnal, they usually occupy dense habitat where they can be tough to see. We also tend to overestimate populations because we see them in groups. Twenty coatis scattering throught the brush can easily look like 40. If we see a group of 20 coatis, we tend to assume that they must be common everywhere, even though we may have seen the only troop in the area. In some tropical resorts, coatis have become used to handouts and are obnoxiously tame. They are likely safer from predators there, so their numbers can get quite large, leading one to think coatis are common everywhere when, in fact, they may be rare outside of the resort.

Impacts of Habitat Changes

Coatis are sensitive to deforestation, which has affected up to 90% of forested habitats in the American tropics. Loss of tropical forests to clearing for livestock and crops, logging, road building, mining and urbanization continues at an increasing rate, with Brazil at the top of the list in rate of forest loss.[11] Except for cases of urbanization, and land completely cleared for industrial agriculture, coatis may fare a little better than many other tropical animals. Coati numbers decline with decreasing forest cover,[12] but they may survive quite well in secondary forests as long as there is sufficient food in the form of insects and fruits, and enough cover to protect them from predators. Indeed, their willingness to use these habitats may facilitate seed dispersal and reforestation.[13]

The sensitivity of coatis to the removal of cover became clear during my study in southeastern Arizona, where portions of my study area were subject to wildfire and fire-prevention treatments, including prescribed fire and ladder-fuel reduction (where the brush and lower tree limbs are cleared to reduce the chance of a fire reaching the tree crown and killing the tree). One area used both treatments at once. Coatis avoided those areas of their home ranges subjected to fire and fire-prevention treatments, although they had been seen in those areas prior to treatment. Both fire and ladder-fuel reduction stimulate basal sprouting in the Madrean oaks – once the sprouts had reached several feet high, creating enough lateral cover, the coatis returned. Reduction of ladder-fuels also reduces habitat for other animals, such as deer and wild turkey,[14] and should only be carefully applied in areas where it is really needed.

Warming climates may be beneficial to coatis by allowing them to expand their range. On the other hand, drying climates, as predicted for the southwestern U.S., means more frequent, high-severity wildfires. Large wildfires and increasing drought may mean that some areas that are now forest will turn to shrubland or savanna, habitats that are not as beneficial to coatis as forest. Many areas at the northern end of their distribution could become unsuitable if forest cover is lost.

Impacts of the Pet Trade

Coatis have been kept as pets for millennia, but a demand for baby coatis in the pet trade can influence populations.[15] Young coatis, particularly where they are habituated near resort areas, are relatively easy to catch, and many thousands have been captured and sold as pets. Although coatis are popular in the exotic pet trade and sellers can be found on the internet, in general, they make terrible pets. They destroy stuff for a living – digging up the ground, turning over logs, climbing up trees and cliffs with ease. Inside a home, they can readily get in unsecured cabinets, rip up carpets, knock things over, and generally create havoc. They bite and scratch, and even their "love nips" (inhibited bites) can be enough to break human skin. A recent attack on some children left deep gashes that required medical treatment.[16] Very few people are capable of giving a coati a good home, and as a result, many are destroyed, turned over to zoos or wildlife centers, or relegated to an outdoor cage to live a sad, lonely life. Coatis are very social animals and need a lot of interaction with either other coatis and/or humans to be happy. In some states, such as Arizona, is it not legal to have a coati without a special permit, which is nearly impossible to obtain.

Coatis are also capable of transmitting diseases, including rabies, distemper, and Chaga's disease, and carry parasites such as worms, fleas, mites, and ticks. Although some exotic animal veterinarians may vaccinate them against rabies and distemper, the vaccines are considered "off-label," meaning they are not administered in the way the vaccine is licensed for. Therefore, if your pet coati bites someone, it may be treated as if it was not vaccinated. There was a tragic situation in Prescott, Arizona, a few years ago in which a woman was bitten by her pet coati. When she reported the bite at the hospital, the coati was taken from her home and euthanized in order to be tested for rabies.[17] In the right home, they can be very loving and entertaining pets, and I was happy to be able to get to know some of Stan and Linda Rolinski's rescued coatis. But I also saw coatis that were mentally "broken" by improper care in their previous homes.

A disturbing trend has developed among some of the exotic animal breeders in the U.S. They have created coati "puppy farms" by mating female coatis multiple times per year – after taking away the newborn kits to be hand raised. Coatis are very attentive and nurturing mothers; removing their newborn kits to increase profits is cruel. In the wild, coatis very rarely have more than one litter per year. In addition, some breeders remove the coatis' claws and canine teeth before they sell them.[18] If you have to mutilate an animal just to bring it in to your home, should you really have it for a pet? These breeders also provide misinformation about coatis, including poor recommendations for diet, and even listing the incorrect species (e.g., listing "mountain" or "red mountain" for brown-nosed coatis), and in some cases, hybridizing brown-nosed and white-nosed coatis. This becomes especially problematic if they are then dumped into the wild in areas where they could potentially mate with wild coatis.

Impacts of Hunting

Throughout their range in the Americas, there are few restrictions on hunting coatis. They are frequently killed by hunters and poachers, however they are seldom considered a preferred species, either by indigenous or recreational hunters.[19,20] Native villagers in the tropics often practice "garden hunting," where they kill animals coming in to raid their gardens, thus providing meat for the pot and reducing damage from the wildlife.[20–22] This type of hunting, as well as poaching and hunting by colonists, may cause a local decline in the numbers of some game species, including coatis.[23,24]

History of Coati Management in the United States

From their "discovery" until the late 1940s, coatis were protected throughout their range in the U.S. In 1948, this protection was lifted in Arizona, and coatis were reclassified as furbearers with unlimited harvest. That meant they could be hunted with any form of gun or bow-and-arrow, as well as trapped using leg-hold traps or snares, with no limits on how many were taken. Why the sudden change of status? I had a fortuitous conversation with an elderly gentleman who lived near the Ramsey Canyon Preserve, when I was radio-tracking coatis one day. According to the gentleman, in the late 1940's one of his neighbors killed a coati that had been raiding a chicken coop, and the neighbor was later arrested by the local game warden.[25] Apparently, the man sued the Arizona Game & Fish Department (AGFD), but when I checked the records for the county, I could find no record of any lawsuit. My suspicion is that he threatened to sue AGFD, which may have been enough for the Game Commission to remove all protections, even though there was no information on coati population numbers or distribution.

Coati numbers appeared to remain relatively stable, possibly increasing, for the next decade. Beginning in the late 1950's, hunters were encouraged to kill coatis because they were blamed for destroying nests of wild turkey and Mearn's quail, and occasionally injured dogs used to hunt mountain lions.[26,27] There was little actual evidence that coatis were destroying nests at all, just an assumption based on their terrestrial foraging habits.[28-30] One resident of the Huachucas recalled to me gunning down entire troops during this period.[31] The impacts of the unlimited hunting and an outbreak of distemper started taking their toll, with a decline in sightings noted in the early 1960s.[27] Indeed, research by a University of Arizona student was thwarted by a lack of coatis to study.[32]

But their status as a furbearer remained unchanged, and an article in *Arizona Highways* magazine in 1962 portrayed them as a horde of voracious Mexican immigrants destroying everything in their path, especially nesting game birds.[26] The only studies of coatis at the time had found no evidence that they consumed anything but fruits and insects.[33] Coatis remained subject to unlimited harvest until 1980 when they were reclassified as a non-game mammal, and take with leg-hold traps was prohibited, although they could still be hunted year-round in unlimited numbers. In 1986, the season was limited to 1 August to 31 March and in 1988, a limit of one animal per calendar year was imposed. Trapping as a method of take was also reinstituted. Arizona banned the use of leg-hold traps on public lands in the mid-1990s, so few coatis are caught in traps anymore. These regulations remain in effect, although coatis are now listed under "Other birds and mammals," which appears to be a catch-all for pests and varmints, including House Sparrows, European Starlings, crows, Gunnison's prairie dogs and most rodents.[34]

Although the term "non-game animal" may be perceived by the public to mean non-hunted or protected from hunting, in Arizona the term refers to *"all wildlife except game mammals, game birds, fur-bearing mammals, predatory mammals, and aquatic wildlife."*[35] The designation includes almost three dozen species of special conservation concern, including species such as the Mexican gray wolf, ocelot, as well as some species of bats, reptiles, and amphibians. Of the non-game mammals, only Gunnison's prairie dogs and coatis may be hunted. No special permits are needed to kill unprotected "other birds and mammals" beyond a standard hunting license, nor are there any attempts to estimate their numbers or actively manage to control or augment their populations. Hunters do not have to report any coatis they kill. Coati cousins, raccoons and ringtails, are considered furbearers, subject to both trapping and hunting, and their harvest numbers are tracked by AGFD.

I repeatedly inquired to AGFD as to why coatis were hunted in Arizona, and the only answer I received was, "*to provide recreational opportunity.*" Over time, and speaking with many hunters, it became clear that hunters seldom sought coatis as prey, but rather, because no special permits were needed, they killed coatis when the opportunity presented itself when hunting more sought-after game such as deer or javelina, or when they were calling in predators such as bobcats or coyotes.

Do hunters in Arizona need more "recreational opportunity?" I combed through the hunting regulations for the 2014-2015 hunting seasons. The minimum number of species that could be hunted on any given day in Arizona was 13 during June, increasing to 41 during November. This understates the number of species, as it lumps together three species of cottontails, two species of jackrabbits, six species of ground squirrels, four species of tree squirrels, and four species of skunks. Some of these, including mountain lions, coyotes, cottontails, jackrabbits, and skunks, can be hunted year-round.

Coati hunting in Arizona appears to be very opportunistic, and at relatively low levels. From the mid-1970s through the late 1990s, taxidermists reported mounting fewer than 14 each year.[36] Although I am unaware of any hunters consuming coatis, the number mounted by taxidermists is likely an underestimate, as I have heard of coatis being shot and left in the field. At current levels, hunting probably has little effect on overall coati populations, however, as mentioned earlier in this book, small troops are exceedingly vulnerable, so if an adult female was hunted out of a troop having only a few adult females, there is an increased chance of the remaining animals succumbing to predators, or abandoning their range to join other groups of coatis. In addition, some of the knowledge of home range, food habits, and other aspects of being a coati that are passed from mother to offspring will be lost, and it can take many years for areas to be recolonized.

As part of an unrelated conversation I had with an AGFD biologist, he asked, "*Why shouldn't we hunt coatis, because they're cute?*" But that is entirely the wrong question. When it comes to taking the life of an animal, any animal, the question is not, "why not?" but rather, "why?" The AGFD biologist's question points out some issues with current game management policies toward non-game species. AGFD, and other state game agencies, proclaim that they manage based on sound scientific principles, and adhere to the tenets of the North American Model of Wildlife Management. But for hunted non-game animals in the state, there are no data on population numbers or trends, no data on the impacts of hunting, no goals or objectives other than providing recreational opportunity to a small proportion of the population that likes to hunt. Providing recreational opportunity cannot be the sole reason for allowing a hunt on any animal.

North American Model of Wildlife Management

The North American Model of Wildlife Management is a set of seven tenets, based on a century-old philosophy of democratizing hunting and developing conservation principles to reduce dramatic losses in wildlife due to market hunting.[37,38] Although not a legal framework, these tenets are held in high esteem by wildlife professionals in state wildlife agencies and commissions.[39] The model has two basic principles – that our wildlife belong to everyone (and are therefore owned by no one), and they should be managed, using scientific principles, for long-term sustainability. Although the model is touted by its authors as one of the greatest conservation success stories in the world, it is not without its detractors.[40] Indeed, as I read through the literature formulating and describing the model, I noticed the glaring absence of two important terms: ecosystem and biodiversity.

The American form of wildlife management described in the model differs from some European forms of management in which hunting is restricted to royalty and/or landowners. This form of management has been successful in recovering numerous game species from the brink of extinction. However, it has occasionally done so at the cost of biodiversity and healthy ecosystems.[41-43] This has been due, in part, both to the way state wildlife agencies are funded, and an overemphasis on big game numbers at the expense of their predators and other carnivorans. The adherence to scientific principles as a tenet of the Model has also been questioned, as management plans, including measurable objectives, transparency, and review processes, have not been developed for most hunted species.[44]

The North American Model of Wildlife Management.[38]

1. Wildlife resources are a Public Trust
Wildlife is owned by no one and is held in trust for future generations.

2. Markets for game are eliminated
No more legal trafficking in meat, parts, and products of game animals and birds (except for fur).

3. Allocation of wildlife is by law
Access to wildlife is allocated to the public by law, not by privilege, and limits are placed on the number of animals harvested to create sustainable populations.

4. Wildlife can be killed only for a legitimate purpose
Legitimate purposes include food, fur, self-defense, or property protection, and should adhere to the sportsmen's code of fair chase and use without waste.

5. Wildlife is considered an international resource
Management of migratory animals needs to be a collaborative effort.

6. Science is the proper tool to discharge wildlife policy
Management of wildlife resources should reflect sound science on wildlife populations, as well as the role of human dimensions.

7. Democracy of hunting is standard
All citizens can participate in hunting and conservation.

Arizona has its own take on the North American Model, which it refers to as Core Concepts.[45] Perhaps recognizing that "providing recreational opportunity" is not consistent with the legitimate use tenet, AGFD has removed that tenet from their Core Concepts. The killing of some of the species listed as non-game or "other" by AGFD could be questioned for their legitimacy. Even The Wildlife Society, a professional organization of wildlife biologists and managers, questioned these types of hunts in their position paper on the North American Model:

> *The current examples of broad-scale prairie dog (Cynomys spp.) shooting and crow hunting raise the question of legitimate purpose. Reconciling this practice within the principle of legitimate use does not seem possible, given that no food or protective benefits are derived.*[38]

Many state wildlife agencies, including AGFD, are funded primarily through the sales of hunting and fishing licenses and matching federal

Arizona's Core Concepts, derived from the North American Model of Wildlife Management.[45]

1. **Wildlife is held in the Public Trust.**
   The public trust doctrine means that wildlife belongs to everyone. Through shared ownership and responsibility, opportunity is provided to all.
2. **Regulated commerce in wildlife.**
   Early laws banning commercial hunting and the sale of meat and hides ensure sustainability through regulation of harvest and regulating commerce of wildlife parts.
3. **Hunting and angling opportunity for all.**
   Opportunity to participate in hunting, angling, and wildlife conservation is guaranteed for all in good standing, not by social status or privilege, financial capacity or land ownership. This concept ensures a broad base of financial support and advocacy for research, monitoring, habitat conservation and law enforcement.
4. **Hunting and angling laws are created through public process.**
   Hunting seasons, harvest limits and penalties imposed for violations are established through laws and regulations. Everyone has the opportunity to shape the laws and regulations applied in wildlife conservation.
5. **Hunters, anglers, boaters and shooters fund conservation.**
   Hunting and fishing license sales and excise taxes on hunting, shooting and fishing equipment and motor boat fuels pay for the management of all wildlife, including wildlife species that are not hunted.
6. **Wildlife is an international resource.**
   Proper stewardship of wildlife and habitats is both a source of national pride and an opportunity to cooperate with other nations with whom we share natural resources. Cooperative management of migrating wildlife is one example of successful international collaboration.
7. **Science is the basis for wildlife policy.**
   The limited use of wildlife as a renewable natural resource is based on sound science. We learn as we go, adapting our management strategies based on monitoring to achieve sustainability.

excise taxes on firearms, ammunition, and fishing gear. Special tags on big game animals – elk, deer, bighorn sheep, pronghorn (antelope), turkey, bear, javelina, bison (buffalo), and sandhill cranes – bring in even more revenue. Lottery systems for big game tags available in some states raise considerable money, and auction tags available through trophy hunting organizations like Safari Club International routinely going for hundreds of thousands of dollars. This creates a huge incentive for state wildlife agencies to manage for trophy-sized animals, and not diverse, balanced ecosystems.

AGFD also receives millions of dollars each year from a portion of state lottery funds (Heritage Funds – which supplied much of the funding for my coati and skunk studies); a non-game checkoff from state income tax revenues; and a portion of tribal gaming revenues. These monies are used for protecting and reintroducing Endangered species, acquiring habitats, wildlife research, and providing wildlife educational opportunities for the general public.[46] So while a majority of their funding comes from hunters and anglers, a sizable portion also comes from the general public. So, in spite of what I consider to be some misguided policies toward some predators and non-game species, AGFD deserves much credit for their education and research programs, reintroduction efforts for endangered species, and other outreach efforts for the general public.

This is not the case in some neighboring states which rely almost exclusively on hunting and fishing licenses and taxes on guns, ammunition, and fishing gear, and as the number of hunters declines each year, so does their revenue. Because most of their income comes from hunting and fishing, state game agencies tend to view those groups as their primary clientele. This, despite their charge to manage all wildlife as part of the public trust, including for those who do not hunt or fish, often referred to as non-consumptive users.

Hunting coatis, in addition to reducing their numbers and potentially causing some troops to abandon their home ranges, makes them very shy toward people. Even in the southeastern quarter of Arizona where they are most common, seeing a coati is an unusual event. Often, sightings are limited to coatis disappearing through the brush as they try to get as far away as possible. I was contacted many times, when I had radiocollared coatis, by people who hoped I could help them get a glimpse of a coati. Coatis are fun to watch and a potential tourist attraction. Hunting coatis, even at the low level practiced currently, precludes many people from being able to enjoy watching these curious animals, and is thus contradictory to the tenets of wildlife as a public trust and hunting only for a legitimate purpose. Because hunters rarely, if ever, buy hunting licenses to specifically hunt coatis, removal of coatis from the list of huntable animals will not affect AGFD's budget. Individual animals can provide much more economic benefit alive than dead.[47] My thoughts echo those of Kaufmann and colleagues from 45 years ago:

> Coatis are unusual in being conspicuous, social, diurnal members of the Carnivora, and they are both amusing and instructive to observe. They are therefore great favorites with the increasingly wildlife- and ecology-conscious public, and should be accorded full protection from hunting and trapping as a valuable and scarce natural resource. Only known individuals which become nuisances on private land or in public campgrounds should be live-trapped and removed, and this potential problem can be reduced by discouraging artificial feeding. Coati numbers in the United States will likely continue to fluctuate with or without any attempts to "manage" them, and we recommend against any such futile and wasteful program on state or federal lands.[10]

Coatis in New Mexico are listed as furbearers with no take allowed. Coatis appear to be increasing in numbers and distribution in New Mexico.[9] Coatis were at the center of a brouhaha in southwestern New Mexico several years ago, when a coati was captured in a leg-hold trap set for bobcats. A hiker passing by heard the coati thrashing in the bushes and crying in pain. He thought he was being merciful when he bludgeoned the coati with rocks to put it out of its misery, only to find himself in trouble with the law for killing a protected furbearer.[48] This incited a lively debate about wildlife trapping in the state, including calling for lifting of protections to allow hunting of "*one of the area's most dangerous critters.*"[49] Although trappers insist that coatis captured in leg-hold traps can be easily released, having handled more than 100 live-trapped coatis, I disagree. A cornered coati is a ferocious beast, and quite willing to use teeth and claws for self-defense. Kaufmann described to me seeing "piles" of coati carcasses discarded by trappers in the 1970s who found it easier to dispatch and dump rather than release.[50]

In Texas, coatis are considered Threatened, and are on the state watch list for Endangered species.[51] Although there are periodic sightings of coatis in Texas, there are no known populations.[52] Coatis receive no protections in Mexico or throughout much of Central or South America.[1]

Coatis fulfill ecological roles – as pollinators and seed dispersers,[13,24,53,54] as prey for big cats (see Chapter 7), as a host for parasites,[55] and in as-yet unstudied roles as soil aerators and insect predators. Current trends in conservation and wildlife biology are beginning to recognize that wildlife has its own intrinsic value,[56] that animals are sentient beings with emotions and capable of suffering, and do not exist solely to fulfill human needs and desires.

Whether or not you agree that nature has its own intrinsic value, or that individual animals matter, the recent furor over the killing of Cecil the lion by an irresponsible trophy hunter made it clear that the general public is not in favor of using animals as target practice.[57] Costa Rica recently banned sport hunting,[58] and countries such as Panama and Peru are limiting hunting by non-indigenous peoples. States have begun banning coyote hunting contests,[59] and even Arizona has banned hunting contests on predators and fur-bearers[60] (alas, this does not apply to coatis). Unfortunately, many wildlife managers take a "with us or against us" attitude that relegates anyone opposed to any form of hunting as an animal rights activist. This precludes discussions of a more scientifically-based and compassionate form of management, which includes some forms of hunting but might limit illegitimate take.

Wild animals are complex and emotional beings, and many have strong family relationships. They all have their place within incredibly complex ecological webs – as prey and predator, as carnivore and scavenger. Due to increasing pressure from ever-increasing human populations, ecosystems around the globe are in danger of unraveling. It is time to re-examine our relationship with the natural world and end destructive "recreational opportunities."

Notes and References

1. Cuaron AD, Helgen K, Reid F, Pino J, Gonzalez-Maya JF (2016) *Nasua narica*: The IUCN Red List of Threatened Species. https://doi.org/10.2305/IUCN.UK.2016-1.RLTS.T41683A45216060.en

2. McFadden KW, García-Vasco D, Cuarón AD, Valenzuela-Galván D, Medellín RA, Gompper ME (2010) Vulnerable island carnivores: the endangered endemic dwarf procyonids from Cozumel Island. Biodiversity and Conservation 19:491–502

3. Emmons L, Helgen K (2016) *Nasua nasua*: The IUCN Red List of Threatened Species. https://doi.org/10.2305/IUCN.UK.2016-1.RLTS.T41684A45216227.en

4. Gonzalez-Maya JF, Reid F, Helgen K (2016) *Nasuella olivacea:* The IUCN Red List of Threatened Species. https://doi.org/10.2305/IUCN.UK.2016-1.RLTS.T72261737A45201571.en

5. Pacheco V, Salas E, Cairampoma L, Noblecilla M, Quintana H, Ortiz F, Palermo P, Ledesma R (2007) Contribución al conocimiento de la diversidad y conservación de los mamíferos en la cuenca del río Apurímac, Perú. Revista Peruana de Biología 14:169–180

6. Pacheco V, Cadenillas R, Salas E, Tello C, Zeballos H (2009) Diversidad y endemismo de los mamíferos del Perú. Revista Peruana de Biología 16:5–32

7. Gonzalez-Maya JF, Arias-Alzate AAA (2016) *Nasuella meridensis:* The IUCN Red List of Threatened Species. https://doi.org/10.2305/IUCN.UK.2016-1.RLTS.T72261777A72261787.en

8. Convention on International Trade in Endangered Species of Wild Fauna and Flora (2013) The CITES Appendices | CITES. https://cites.org/eng/app/index.php. Accessed 26 Mar 2016

9. Frey J, Lewis J, Guy R, Stuart J (2013) Use of anecdotal occurrence data in species distribution models: an example based on the white-nosed coati (*Nasua narica*) in the American Southwest. Animals 3:327–348

10. Kaufmann JH, Lanning DV, Poole S (1976) Current status and distribution of the coati in the United States. Journal of Mammalogy 57:621–637

11. Lindsey R (2007) Tropical Deforestation: Feature Articles. *In:* Earth Observatory. http://earthobservatory.nasa.gov/Features/Deforestation/. Accessed 11 Apr 2016

12. Urquiza-Haas T, Peres CA, Dolman PM (2011) Large vertebrate responses to forest cover and hunting pressure in communal landholdings and protected areas of the Yucatan Peninsula, Mexico. Animal Conservation 14:271–282

13. Alves-Costa CP, Eterovick PC (2007) Seed dispersal services by coatis (*Nasua nasua*, Procyonidae) and their redundancy with other frugivores in southeastern Brazil. Acta Oecologica 32:77–92

14. Personal observation.

15. Leopold AS (1959) Wildlife of Mexico. The game birds and mammals. University of California Press, Berkeley

16. Bittner GC, Hans NR, Neto GH, Morais MO, Filho GH, Haddad VJr (2010) Coati (*Nasua nasua*) attacks on humans: case report. Wilderness & Environmental Medicine 21:349–352

17. Anonymous (2012) Prescott Valley woman bitten by pet coati. *In:* AZFamily.com. http://www.azfamily.com/story/28347191/prescott-valley-woman-bitten-by-pet-coati. Accessed 29 Mar 2016

18. Castleberry R (2010) Raising and caring for your pet coatimundi. Tifkar Publishing, Lampass, TX

19. Jorgenson JP, Redford KH (1993) Humans and big cats as predators in the Neotropics. Symposium of the Zoological Society of London 65:367–390

20. Redford KH, Robinson JG (1987) The game of choice: patterns of Indian and colonist hunting in the Neotropics. American Anthropologist 89:650–667

21. Escamilla A, Sanvicente M, Sosa M, Galindo-Leal C (2000) Habitat mosaic, wildlife availability, and hunting in the tropical forest of Calakmul, Mexico. Conservation Biology 14:1592–1601

22. Vickers WT (1991) Hunting yields and game composition over ten years in an Amazon Indian territory. *In*: Robinson JG (ed) Neotropical wildlife use and conservation. University of Chicago Press, Chicago, IL, pp 53–81

23. Glanz WE (1991) Mammalian densities at protected versus hunted sites in central Panama. *In*: Robinson JG (ed) Neotropical wildlife use and conservation. University of Chicago Press, Chicago, IL, pp 163–173

24. Wright SJ, Zeballos H, Domínguez I, Gallardo MM, Moreno MC, Ibáñez R (2000) Poachers alter mammal abundance, seed dispersal, and seed predation in a Neotropical forest. Conservation Biology 14:227–239

25. Fred Thomas, personal communication.

26. Healy IH (1952) The coati mundi. Arizona Highways May 1952:31–33

27. Pratt JJ (1962) Establishment and trends of coati-mundi in the Huachucas. Modern Game Breeding 23:10–11, 15

28. Coatis introduced onto islands have negatively impacted sea bird populations (Jaksic 1998), and they are known to consume sea turtle eggs in Costa Rica (Fowler 1979), but neither of these facts were available to wildlife managers in Arizona in the 1950s.

29. Jaksic FM (1998) Vertebrate invaders and their ecological impacts in Chile. Biodiversity and Conservation 7:1427–1445

30. Fowler LE (1979) Hatching success and nest predation in the green sea turtle, *Chelonia mydas*, at Tortuguero, Costa Rica. Ecology 60:946–955

31. Tom Beatty, personal communication.

32. Risser SC Jr (1963) A study of the coati mundi *Nasua narica* in southern Arizona. Master's Thesis, University of Arizona, Tucson

33. Wallmo OC, Gallizioli S (1954) Status of the coati in Arizona. Journal of Mammalogy 35:48–54

34. Arizona Game & Fish Department (2015) Hunt Arizona 2015 Edition.

35. 1995-1996 Arizona Hunting Regulations, Arizona Game & Fish Dept., Phoenix, AZ.

36. Hass CC (1997) Ecology of white-nosed coatis in the Huachuca Mountains, Arizona, a preliminary study. Final report submitted to the Arizona Game & Fish Department, Phoenix, AZ. 1-52

37. Geist V, Mahoney SP, Organ JF (2001) Why hunting has defined the North American Model of Wildlife Conservation. *In*: Transactions of the North American Wildlife and Natural Resources Conference. Wildlife Management Institute, Washington, D.C., pp 175–185

38. Organ JF, Geist V, Mahoney SP, et al (2012) The North American Model of Wildlife Management. The Wildlife Society, Bethesda, MD

39. Vucetich J (2013) Dr. John Vucetich: On Hunting Wolves. *In*: Between Thorns and Claws. https://4thenaturesake.wordpress.com/2013/06/05/dr-john-vucetich-on-hunting-wolves/ Accessed 29 Apr 2018

40. Nelson MP, Vucetich JA, Paquet PC, Bump JK (2011) An inadequate construct? North American model: What's flawed, what's missing, what's needed. Wildlife Professional Summer 2011:58–60

41. Eisenberg C (2010) The wolf's tooth: keystone predators, trophic cascades, and biodiversity. Island Press, Washington, D.C.

42. Fraser C (2011) The crucial role of predators: a new perspective on ecology. *In*: Yale Environment 360. http://e360.yale.edu/feature/the_crucial_role_of_predators_a_new_perspective_on_ecology/2442/ Accessed 9 Apr 2016

43. Hass CC (2009) Competition and coexistence in sympatric bobcats and pumas. Journal of Zoology 278:174–180

44. Artelle KA, Reynolds JD, Treves A, Walsh JC, Paquet PC, Darimont CT (2018) Hallmarks of science missing from North American wildlife management. Science Advances 4:eaao0167

45. https://www.azgfd.com/hunting/nam/ Accessed 12 Dec 2020

46. https://www.azgfd.com/Wildlife/HeritageFund/Program/ Accessed 12 Dec 2020

47. Elbroch LM, Robertson L, Combs K, Fitzgerald J (2017) Contrasting bobcat values. Biodiversity and Conservation 26:2987–2992

48. Fryxell DA (2005) The quality of mercy. *In*: Desert Exposure. http://www.desertexposure.com/200506/200506_mercy.html. Accessed 1 Jan 2016

49. Lightner L (2010) The fiercest of beasts. In: Desert Exposure. http://www.desertexposure.com/201005/201005_outdoors.php. Accessed 27 Mar 2016

50. John Kaufmann, personal communication.

51. Schmidly DJ (2004) The mammals of Texas, Revised. University of Texas Press, Austin, TX

52. Jonah Evans, Texas state mammologist, personal communication.

53. Mora JM, Mendez VV, Gomez LD (1999) White-nosed coati *Nasua narica* (Carnivora: Procyonidae) as a potential pollinator of *Ochroma pyramidale* (Bombaceae). Revista de Biologia Tropicale 47:719–721

54. Sáenz JM (1994) Ecología del pizote (*Nasua narica*) y su papel como dispersador de semillas en el bosque seco tropical, Costa Rica. Master's Thesis, Universidad Nacional, Heredia, Costa Rica

55. Gompper ME (2004) Correlations of coati (*Nasua narica*) social structure with parasitism by ticks and chiggers. *In*: Sanchez-Cordero V, Medellin RA (eds) Contribuciones mastologolicas en homenaje a Bernardo Villa. Instituto de Biologia e Instituto de Ecología, UNAM, Mexico City, pp 527–534

56. Vucetich JA, Bruskotter JT, Nelson MP (2015) Evaluating whether nature's intrinsic value is an axiom of or anathema to conservation. Conservation Biology 29:321–332

57. Bekoff M (2015) Compassionate conservation, Cecil the murdered lion, and Blaze the slaughtered Yellowstone bear. *In*: The Huffington Post. http://www.huffingtonpost.com/marc-bekoff/compassionate-conservatio_2_b_8003676.html. Accessed 11 Apr 2016

58. Zimmer L (2015) Costa Rica becomes first Latin American country to ban hunting for sport. http://inhabitat.com/costa-rica-becomes-first-latin-american-country-to-ban-hunting-for-sport/ Accessed 11 Apr 2016

59. Rott N (2014) California bans coyote killing contests. https://www.npr.org/2014/12/04/368408213/california-bans-coyote-killing-contests. Accessed 5 Dec 2018

60. Arizona Game & Fish Department (2019) Rule prohibiting organized predator hunting contests is in effect. https://www.azgfd.com/wildlife/nongamemanagement/ Accessed 15 Dec 2019

Acknowledgments

This project would not have been possible without the support of many people and organizations. I am hugely grateful to Sheridan Stone, then Wildlife Biologist at Fort Huachuca, for answering my queries, facilitating funding and access to the Fort, proofreading numerous proposals and manuscripts, and sharing observations and ideas. Funding for the study was provided by Arizona Game & Fish Heritage Funds, with considerable in-kind support from the Fort Huachuca Wildlife Office. Marty Tuegel provided additional financial and logistical support, including designing equipment and field assistance. Dan Pond also helped design field equipment and advised on collecting biological samples. Chris Kochanny, then with Advanced Telemetry Systems, designed and produced the radio collars.

For the first year and a half of the study, I had a part-time research assistant. Jason Roback assisted in handling animals, but mostly was assigned to radio-tracking. Young and agile, he had few qualms about bushwhacking through skin-shredding brush and cactus to get a glimpse of a troop of coatis. He worked long hours for little pay, but his sense of humor, field skills, and ability to take good field notes were crucial to the study. I was also extremely fortunate that Dr. John (Jack) Kaufmann had just retired from the University of Florida and opted to spend a couple of winters in southeastern Arizona. He frequently drove over from his rental cabin in the Chiricahua Mountains to assist with radio-tracking. His willingness to share his experiences from many years of studying coatis was invaluable. Mike Seidman, then Director of Conservation at the Phoenix Zoo, was with the study the longest. He assisted with radio-tracking and hunting down sources of information from the early site visits in 1994 until 2000. I am also indebted to countless hikers and birders I ran into over the years that shared stories of their encounters with coatis.

The Nature Conservancy staff, including Tom Wood, Paul Hardy, Mark Pretti, and others, graciously allowed access to the Ramsey Canyon Preserve for observing and trapping coatis, and shared observations. Mark Fredlake, Bureau of Land Management, facilitated access to the San Pedro Riparian National Conservation Area for trapping and field surveys. Michael Sutor provided valuable observations and graciously shared video footage and audio recordings of coatis. I thank Melanie Bucci and Yar Petryszyn for allowing access to the mammal collections at the University of Arizona, Shawnee Riplog-Peterson for facilitating access to the captive coatis at the Arizona-Sonora Desert Museum and sharing their individual histories, and Megan Pitman and Travis Perry for sharing photos from their carnivore research in New Mexico.

Stan and Linda Rolinski let me into the lives of their rescued coatis, first in Arizona and then in Florida, letting me film them, and later collecting sound recordings for me. The differences in behavior of captive coatis and wild coatis let me see how truly flexible these animals are. Conversations with other biologists studying coatis also provided insight into how these animals adapt to different environments, and I am grateful to David Valenzuela, Ben Hirsch, Matt Gompper, Mirian Tsuchiya, and Corina Logan.

I thank Janice Przybyl for assisting me with filming coatis during sound playbacks at the Arizona-Sonora Desert Museum and accompanying me on rock art adventures. I am also grateful to Glenn Omundsen for sharing some fabulous rock art sites with me in Texas, New Mexico, and Arizona. Mike Foster graciously shared video of coatis, many stories, and accompanied me on trips to the San Pedro River and to rock art sites in Mexico. David Valenzuela kindly supplied videotapes and some hard-to-find references. Jerry Dragoo assisted with the phylogeny. Emily Grout appeared during the final stages of this project, and was very helpful in sorting out coati vocalizations and starting to make sense of coati language.

Long discussions with Kevin Hansen, during the writing of his book *Bobcat: master of survival*, were helpful in shaping the direction of this book. The manuscript benefited from discussions, reviews, and comments by Harley Shaw, Jason Roback, Glenn Kellogg, Joe Letsche, Emily Grout and Ben Hirsch. Of course, final errors and omissions are mine.

Appendix I. Common and scientific names of plants and animals

Plants

Acacia, catclaw	*Senegalia greggii*
Agave	
Palmer's	*Agave palmeri*
Parry's	*Agave parryi*
Balsa	*Ochroma pyramidale*
Barberry, Wilcox	*Berberis wilcoxii*
Buckthorn, birch-leafed	*Frangula betufolia*
Chokecherry	*Prunus virginiana*
Corn (Maize)	*Zea mays*
Cottonwood, Fremont	*Populus fremontii*
Francincense	*Boswellia* spp.
Grape, canyon	*Vitis arizonica*
Hackberry, desert	*Celtis pallida*
Juniper, alligator (alligator-bark)	*Juniperus deppeana*
Madrone, Arizona	*Arbutus arizonica*
Manzanita, pointleaf	*Arctostaphylos pungens*
Maple, bigtooth	*Acer grandidentatum*
Myrrh	*Commiphora* spp.
Oak, silver-leafed	*Quercus hypoluecoides*
Papaya	*Carica papaya*
Pinyon, Mexican	*Pinus cembroides*
Prickly Pear, Englemann's	*Opuntia phaaeacantha*
Raspberry	*Rubus* spp.
Sumac, Mearn's	*Rhus choriophylla*
Sycamore, Arizona	*Platanus wrightii*
Walnut, Arizona	*Juglans major*

Birds

Eagle, golden	*Aquila chrysaetos*

Owls
  Mexican spotted   *Strix occidentalis*
  Whiskered screech-   *Megascops trichopsis*
Quail
  California   *Callipepla californica*
  Mearn's   *Cyrtonyx montezuma*
Robin, American   *Turdus migratorius*
Sparrow, song   *Melospiza melodia*
Titmouse, bridled   *Baeolophus wollweberi*
Turkey, wild   *Meleagris gallopavo*

Mammals
Badger, American   *Taxidea taxus*
Bear
  black   *Ursus americanus*
  grizzly   *Ursus arctos*
  polar   *Ursus maritimus*
Bobcat   *Lynx rufus*
Coyote   *Canis latrans*
Deer
  mule   *Odocoileus hemionus*
  white-tailed   *Odocoileus virginiana*
Fox
  European   *Vulpes vulpes*
  gray   *Urocyon cinereoargenteus*
Elephant   *Loxodonta africana*
Hippopotamus   *Hippopotamus amphibious*
Hyena, spotted   *Crocuta crocuta*
Jackal, silver-backed   *Canis mesomelas*
Jackrabbit   *Lepus* spp.
Jaguar   *Panthera onca*
Javelina (collared peccary) *Pecari tajacu*
Leopard   *Panthera pardus*
Lion
  African   *Panthera leo*
  mountain   *Puma concolor*
Margay   *Leopardalis wiedii*
Marmots   *Marmota* spp.

Meerkat	*Suricata suricata*
Mongoose	
banded	*Mungos mungo*
yellow	*Cynictis penicillate*
Monkey	
vervet	*Chloracebus pygerythrus*
capuchin	*Cebus imitator*
Ocelot	*Leopardalis pardalis*
Opossum, Virginia	*Didelphis virginiana*
Panda, giant	*Ailuropoda melanoleuca*
Rat, pack	*Neotoma* spp.
Sheep, bighorn	*Ovis canadensis*
Skunks	
hog-nosed	*Conepatus leuconotus*
Hooded	*Mephitis macroura*
Striped	*Mephitis mephitis*
western spotted	*Spilogale gracilis*
Squirrel	
Arizona gray	*Sciurus arizonensis*
California ground	*Otospermophilus beecheyi*
Richardson's ground	*Urocitellus richardsonii*
rock	*Otospermophilus variegatus*
Tayra	*Eira Barbara*
Tiger	*Panthera tigris*
Walrus	*Odobenus rosmarus*
Weasel, least	*Mustela nivalis*
Wolf, gray	*Canis lupus*

Reptiles

Constrictor, boa	*Boa constrictor*
Rattlesnake, black-tailed	*Crotalus molossus*

Invertebrates

Centipede, desert	*Scolopendra heros*

Index

Agave 25, 105, 165, 170–171

Ailuridae 8

Allee effect 104, 108

Aravaipa Creek 170, 190

Arctonasua 215

Argentina 2, 9, 11, 37, 44, 51, 55, 57, 59, 69, 74, 76–77, 84, 87, 173, 183, 204

Arizona Game & Fish Department (AGFD) 235, 237, 239, 241

Arizona-Sonora Desert Museum 120, 124, 134, 147, 159

Aureli, Teresa 72

Bassaricyon 10
  neblina 11
Bassariscus 10
  astutus 10
  sumacrasti 10
Bear 1, 3, 8, 165
  black 3, 5, 45, 86, 168, 172, 182, 240
  brown 126
  polar 3
Behavior 32, 37–38, 43, 116, 154, 179. See Also Social behavior
  anti-predator 90, 157, 161
  captivity 78, 120
  decoy 45–46
  denning 76, 165, 188
  foraging 50, 73–74, 83, 89, 106–107, 115, 131–135, 140–141, 147, 154, 160, 165, 172–173, 175, 206

Binczik, Gerald 56, 58–62

Birth 43, 68, 87
  age at first reproduction 47
  emergence from nest 49, 160
  gestation 55, 57, 200
  litter size 47, 160
  nests 11, 44–45, 47–49, 103, 188
  seasonality 3, 48, 55–58
  synchronicity 48–49, 53, 58–62, 200
  weight 44, 52
Boa constrictor 85–86
Bobcat 34, 87, 172, 237, 242
Booth-Binczik, Susan 76, 83, 87, 95, 201, 204
Bowers, Janice 20
Burseraceae 137
Byers, John 197

Cacomistle 10
Caniformia 8
Capuchin monkeys 47, 88, 173
Caso, Arturo 76
CDC (Centers for Disease Control) 100, 102
Childs, Jack 102
Chiricahua Mountains 76, 104, 182, 185, 190
CITES (Convention on International Trade in Endangered Species) 232
Clutton-Brock, Tim 92

Coati
  archaeological evidence 216–
  217, 224, 226
  as metaphor 219, 224, 226
  as pets 120, 213, 217–218, 224,
  234–235
  as seed dispersers 169, 233
  bite force 173
  body size 2, 49, 78, 119
  capturing 32, 35, 104
  coalitions 79, 84, 90, 92, 145,
  150
  diet 3–4, 38, 56, 84, 126, 166,
  169–170, 173–174, 190, 235, 237
  disease 99, 103, 184, 232

    distemper 69, 99, 184
    rabies 99, 103, 234
  distribution 6, 11, 224, 226,
  232, 234, 236, 242
  evolution 8–10
  frontal cortex 78
  group composition 4, 68–69, 72–
  73, 76, 84, 89–90, 93, 99, 137,
  203
  harvest 235–237
  importance of culture 110
  Isle of 219
  names 4, 6–7
  parasites 91, 99, 103, 111, 136–
  137, 234, 243
  scent glands 125–126
  sexual dimorphism 2
  societies 4, 75
  teeth 1, 4, 47, 73, 85, 136,
  145, 148–149, 173, 200, 235
Cognitive map 180
Coloration
  aposematic 121
  coat 2, 7, 119
  eye spots 2, 121–122
Columbia 2, 10–12, 38, 69, 122,
  232

Communication 115, 119–120,
  145, 157, 179
  acoustic 118, 124, 156, 160
  olfactory 125, 127, 139
  tactile 135
  visual 120–121
Compton, L.A. 133
Convergent evolution 10
Costa Rica 7, 37, 47, 56, 59, 76,
  86, 88, 137, 171, 173, 243
Cozumel Island coati 11, 217–218,
  232

Darwin, Charles 146
Deer 87, 118, 172, 216, 233, 237,
  240
  mule 23, 67, 88, 107
  white-tailed 23, 67, 88, 107, 172

Elephants 70, 111, 116
Encinal 20–21, 23, 26, 174, 190
ENSO (El Niño Southern Oscillation)
  78

Feliformia 8
Fisher, Amy 19
Fort Huachuca 20, 26–27, 31–32,
  34, 36, 38, 68, 106–107, 157,
  174, 190, 214–215
Foxworthy, Jeff 3

GABI (Great American Biotic
  Interchange) 8
Gadsden Purchase 214
Gasco, Aline 124, 131–133, 147–
  148
Gilbert, Bil 76, 84, 123, 199–200
Gittleman, John 4

Gompper, Matthew 72, 75–76, 83, 91, 137–138, 204

Gray fox 34, 99–101, 126, 165, 167–168

Grout, Emily 124

Guatemala 7, 37, 44, 49, 56, 58–61, 68, 76, 87, 132, 140–141, 155, 173, 175, 201, 203–204

Habitat 21, 76, 92, 222, 231
availability 24, 174–175, 203, 214, 232
deforestation 226, 231–233
effects on communication 118–119, 121–122
horizontal cover 191, 232–233
use 169–170, 189–190

Hirsch, Ben 51, 59, 61–62, 69, 72, 74, 76, 79, 84, 87, 93–94, 147, 200–201

Hoffmeister, Donald 7

Holmgren, Virginia 7

Home range 70, 105, 179–180, 184–185, 187, 196, 205–206, 208, 233, 237, 241
calculation 180, 182, 192
importance to survival 186–187, 207
overlap 36, 69, 72, 183–186, 205
size 182–183

Hoylman, Anne 137–138

IUCN (International Union for the Conservation of Nature) 231–232

Jaguar 24, 87, 102, 221

Juniper 219
alligator 21, 84, 115, 167–170, 173

Kaufmann, John 37, 123, 133, 136, 141, 156, 170, 190, 199–201, 204, 214, 242

Kinkajou 3–5, 8, 10, 78, 218

Lanning, Dirk 190

Lemurs
redfronted 132
ringtailed 3, 139

Leopard 156

Lion
African 92, 197
mountain 5, 44–45, 75, 86–90, 105–108, 126, 138, 155, 172, 187–188, 207, 218–221, 236–237

Lutrinae 8

Mating 126, 137, 199
copulation 198, 200, 202
fighting 77, 85, 149, 206
movements 68, 191, 196, 205
paternity 204–205
scent marking 205
season 3, 43, 56–58, 60, 68, 75, 141, 150, 174, 195, 202, 207–208
synchrony 61
system 4, 75, 196, 198–203, 208

Maurello, M.A. 133

McColgin, Maureen 72, 76, 104, 183, 185

McLaughlin, Steve 20

Mephitidae 8

Mexico 7, 11, 23, 25, 27, 37, 56, 59, 76–77, 87, 99, 104, 173, 175, 179, 183, 195, 215–217, 219, 222, 224, 226, 242

Mongoose 8
banded 92
meerkat 91
yellow 154, 157

Monsoon 22, 24, 37, 55, 174, 225–226
  impacts on reproduction 55, 78
Murrah Cave 215

Nasua 7, 10, 12, 121
  mastodonta 215
  narica 7, 215
  nasua 7
  nelsoni 11
  pronarica 215
  sociabilis 7
  solitaris 7
Nasuella 7, 10, 12
  meridensis 12
  olivacea 12

North American Model of Wildlife Management 238–240

Ocelot 85–86, 88, 236
Odobenidae 8
Olingo 8, 10–11
Opossums 34, 91
Otariidae 8
Owl 102, 116
  spotted 86
  whiskered screech- 172

Paleoclimate 224, 226
Panama 7–8, 11, 13, 37, 44, 47, 56, 58–59, 62, 74–76, 83, 85–86, 109–110, 123–124, 137, 173, 199–200, 204, 243
Paranasua 215
Pineal gland 57, 64
Pleistocene 23–24, 217
Population 37, 48, 76, 87, 93, 104, 110, 191, 196–197, 201, 204, 206, 208, 213, 226, 231–232, 234, 236–237, 242
  declines 56, 99, 104–105, 107–108, 184, 226, 232
  density 104, 184, 203
  genetic variation 9, 11, 205
Potos 10
Pratt, Jerry 32
Procyonidae 8–10
  ancestral procyonids 8
Pronghorn 197, 216, 240

Quail
  California 157
  Mearn's 236

Raccoon 3, 34, 117, 126, 170, 215, 217, 237
  bite force 173
  communication 124, 126, 134
  dispersal 111
  fall weight gain 3, 174
  frontal cortex 78
  growth rates 78
  hierarchies 79
  litter weight 44
  paternity 204
  phylogeny 8, 10
  rabies 103
  social behavior 111
  social organization 77, 111
  teeth 4–5
Ramsey Canyon Preserve (TNC) 26, 31, 34, 48, 68, 75, 77, 85, 104, 115, 140, 142, 159, 172, 235
Ratneyeke, Shayamala 38, 183
Rattlesnake 34, 67
Ringtail 3, 5, 8, 10, 31, 33–35, 125–126, 170, 216, 237
Roback, Jason 31–33, 45, 149, 156, 172, 174, 198–200, 202, 220

Index

Rolinski, Stan and Linda 120, 124, 138, 234

Romero, Filippo 72

Russell, Jim 37, 58, 74, 83–85, 93

Russell, Will 215, 218, 222

San Pedro River 24, 26, 34, 170, 189–190

Seidman, Mike 106–107

Senses
hearing 75, 118, 154, 158–159
smell 74, 116–117, 166
taste 116
touch 72, 116–117, 119, 135
vision 10, 116, 119

Sign 31, 106, 165, 187
diggings 31, 126
scat 165–167, 170, 176
tracks 26, 126, 165, 187
tree nests 188–189

Simpson, George Gaylord 7

Skunks 3, 8, 11, 34–35, 91, 99– 103, 237
hog-nosed 34, 102
hooded 24, 31–32, 34, 101
spotted 31
striped 101

Slobodchikoff, Con 127

Smith, Harriet 123, 136

Social behavior 19, 78
affiliative 72, 131, 137, 141
agonistic 110, 131, 145, 206
allogrooming (mutual grooming) 72–73, 91, 96, 123, 131, 135–138
anointing 138
cultural transmission 90–91, 109–111, 139
dominance hierarchy 78–79
greeting 131
maternal 44
play 90, 131, 139

principle of antithesis 146
reconciliation 141

Squirrel 156–157, 170, 237
California ground 119–120
gray 34
rock 34

Stone, Sheridan 20

Sutor, Michael 46, 89, 140

Territory. See Home range

Tigers 121

Trattinnickia aspera 137–139

Treaty of Guadelupe Hidalgo 214

Ursidae 8

Valenzuela, David 32, 87, 104

Van Valkenburgh, Blaire 4

Wild turkey 168, 175, 182, 233, 236

Wolf 216, 236

Woosley, Anne 215

www.ingramcontent.com/pod-product-compliance
Lightning Source LLC
Chambersburg PA
CBHW031545260326
41914CB00002B/284

.